S0-BSM-771

OPERATIONS OF
THE GEOMETRIC
AND MILITARY
COMPASS

Galileus Galileus   Florentinus

Superior   licentia

1 6   2 4.

Eques Octauius Leonus   Roman pictor fecit

GALILEO GALILEI

OPERATIONS OF
THE GEOMETRIC
AND MILITARY
COMPASS
1606

TRANSLATED, WITH AN INTRODUCTION

By

STILLMAN DRAKE

Published for

THE BURNDY LIBRARY

by

THE DIBNER LIBRARY OF THE HISTORY
OF SCIENCE AND TECHNOLOGY

and the

SMITHSONIAN INSTITUTION PRESS
WASHINGTON, D.C.   1978

Illustration Credits

Frontispiece courtesy of the Bern Dibner Collection, Norwalk, Connecticut
Pages 13, 14 (bottom), 16, 17, 27, 28, 30, and 32 courtesy of the
University of Toronto Photographic Service
Page 14 courtesy of the Adler Planetarium, Chicago, Illinois
Page 18 from "Galileo and the First Mechanical Computing Device" by
Stillman Drake. Copyright © 1976 by Scientific American, Inc.
All rights reserved.
Page 21 courtesy of the British Museum Library, London
Page 38 and line drawings on pages 44–91 courtesy of the Library of
Congress, Rosenwald Collection

Cover and Frontispiece

Galileo at age 60, engraved by
Ottavio Leoni in Rome in 1624 during the
scientist's visit to honor the new Pope Urban VIII. From the Bern Dibner
collection, Norwalk, Connecticut.

Designed by Natalie E. Bigelow

Copyright © 1978 by the Dibner Library of the History of Science and Technology.
All rights reserved.

Library of Congress Cataloging in Publication Data

Galilei, Galileo, 1564–1642.
Operations of the geometric and military compass, 1606.
(Publication—Dibner Library; no. 1)
Translation of Le operazioni del compasso geometrico, et militare.
Bibliography: p. 34.
1. Mathematical instruments—Early works to 1800.
2. Geometry—Early works to 1800.
3. Surveyor's compass—Early works to 1800.
I. Title.
II. Series: Dibner Library.
Publication—Dibner Library; no. 1.
QA71.G2713   1978   681'.75   78–606002

BURNDY LIBRARY
PUBLICATION NO. 32

CONTENTS

I shall not say what I have accomplished by this work of mine,
but let those judge who have learned from me up to now, or will
learn in the future . . .

GALILEO GALILEI

1606

Something of the importance to society of such an invention as
Galileo's can be grasped from the modern introduction of the
pocket electronic computer . . . that suddenly made it possible for
nearly everyone to deal effectively with almost any problem arising
in practical matters by following rather simple instructions . . .

STILLMAN DRAKE

1978

INTRODUCTION

I.

Soon after Galileo was appointed to the chair of mathematics at the University of Padua in 1592, he began to give private instruction in certain subjects not covered by the official curriculum. Among the first of those courses to be offered was one on military architecture and fortification. Galileo's "geometric and military compass" grew out of that activity and became the subject of his first acknowledged printed book.

The instrument started out as a consolidation of two others that had been long in general use by artillery men, with some improvements and extensions of their uses. In that first form the only scales marked on it were on the quadrant arc. In 1596 Galileo composed a little treatise on the measurement of distances and altitudes by sighting and triangulation which survives as an appendix in his later book, virtually unchanged except for several brief but significant additions made in 1599. It was in 1597 that Galileo began to mark scales along the arms of the instrument, commencing with two that were subsequently eliminated; those first two he simply took from a drafting instrument that was probably devised not long before by his patron and friend, the Marquis Guidobaldo del Monte. In the same year he added other scales, some of which were of his own invention, that were of great value to artillery-men as well as to military engineers. The earliest form of Galileo's instructions for the use of his "compass" belongs to 1597; of this there are two copies, both formerly owned by another friend and patron, Gian Vincenzo Pinelli, now at the Ambrosiana Library in Milan. At this stage it had become a mechanical calculating device but remained limited in application to various geometric and military problems. By the beginning of 1599 Galileo had developed it into a general-purpose mechanical calculator capable of solving any practical mathematical problem that was likely to arise—swiftly, simply, without requiring previous mathematical education, and sufficiently accurately for ordinary practical purposes.

No previously known instrument had accomplished anything quite like that, although mechanical aids to calculation had appeared in various forms. The oldest was doubtless the abacus, limited to arithmetical computations and of little use for root extraction or even long-division problems. The astrolabe affords an example of special-purpose mechanical computing aids, ingenious and complex but of restricted applicability to ordinary practical mathematics.

Something of the importance to society of such an invention as Galileo's can be grasped from the modern introduction of the pocket electronic computer. Before it there had been slide rules, used mainly in engineering problems and requiring a certain amount of technical mathematical understanding of its operations. There were also desk calculators, neither inexpensive nor portable, and (like the abacus) not very efficient in root extraction, which is required in the solution of many geometric and algebraic problems. It was the pocket electronic computer that suddenly made it possible for nearly everyone to deal effectively with almost any problem arising in practical matters by following rather simple instructions. The effects on society are already noticeable after only a few years.

It was something like that that happened early in the seventeenth century with the rapid spread of Galileo's "compass," properly known in English as the "sector." This name was conferred on a simpler but very similar instrument in 1598 by Thomas Hood, who published a book on it at London entirely independently of Galileo's work. One of the immediate consequences was that topographical surveying and mapping of terrain became possible for anyone interested, no longer requiring trained specialists. Thus in 1604 Galileo published (under a pseudonym) a facetious dialogue poking fun at a famous philosopher's notion of the location of a new star, and he made one of the peasant interlocutors speak as a shepherd-turned-surveyor, something hardly conceivable before 1600. The sector probably had prompt effects at the Venetian arsenal, a place Galileo frequently visited to talk with workmen and foremen, as he did with craftsmen and mechanics generally. To work from models to full-scale structures had formerly required the presence of specialists; now the calculating sector made this a simple matter of settings and readings. Wastage of time and material in trial-and-error designing must have been much reduced by the advent of mechanical approximations in proportion problems. Area and volume relations no longer required pencil-and-paper calculations of which few persons were capable. And just as the level of capabilities was raised among workmen, the reliance of princes on estimates by captains and engineers was no longer necessary, since it had become possible for them to check these for themselves without lengthy calculations.

Since Galileo's book explains the uses of the instrument, this introduction is intended mainly to outline the steps in its origin and development and its fate after his book was published. A few general remarks about his sector will suffice before proceeding to those things. The scales were engraved with a precision within about one percent. In practical problems we are seldom concerned with more than two or three significant places, in which respect Galileo's sector was about as good as a modern slide rule. It was furthermore a better instrument for men not trained in mathematics, since on the sector most geometrical problems can be solved without translating them into numbers and then translating the solutions back into geometrical magnitudes. Operations with the sector are carried out with more ease than their verbal descriptions suggest. Basically there are only three operations: setting the separation of the arms, taking the distance from the pivot to a point along one of its scales, and taking the crosswise distance between a point and the corresponding point on the other arm. When necessary these distances are taken on a pair of dividers so that they may be transported with great precision to another position on the sector, or to paper, or to a physical object involved in the problem.

II.

If, like Thomas Hood, Galileo had thought to give his instrument a really distinctive name, like "sector," much historical confusion and misunderstanding that arose later might have been avoided. Some of the trouble has arisen because other writers came to call instruments like Galileo's the "proportional compass" despite the fact that the name had previously been applied, and is still today applied in English, to quite a different instrument. This is very useful in drafting; it has two slotted arms with sharp points at both ends, whence Galileo called it the "compass of four points." He was perfectly familiar with it, and his instrument-maker manufactured some for him and for sale to students, but Galileo never claimed to have invented it.

The proportional compass probably originated at Urbino before 1575, at the suggestion of a man named Bartolommeo Eustachio. In the slots of its arms there is a cursor serving as pivot that can be held wherever desired by means of a screw. This pivot keeps the distances between the two pairs of compass points in the same ratio, whatever their separations. Prior to this invention it appears that draftsmen customarily had sets of X-shaped instruments fixed at ratios of 2:1, 3:1, and so on for their wider and narrower points. Eustachio wanted a single device that would be adjustable to any desired ratio and is said to have asked the mathematician Federico Commandino of Urbino to design such an instrument. The proportional compass ordinarily has two scales of numbers on it, one of which shows the number of times the distance between the narrow pair of points will fit into the distance between the wide pair. The other scale usually gives the number of sides of a regular polygon such that the wide pair of points gives the diameter of the circumscribing circle, while the narrow pair gives the length of one side of the polygon.

The proportional compass is still manufactured today with hardly any difference from those of Galileo's time. In 1604 it was described in a book by Levinus Hulsius, who had been a student of Galileo's at Padua. Hulsius attributed the invention to the Swiss mathematician Joost Burgi, who moved to Prague in 1603 and became a friend of Johann Kepler. It is not in any sense a calculating instrument, nor is the Galilean sector of much use directly in drafting, as is the proportional compass.

Another very precise drafting instrument that appears to have been invented before 1570 was the reduction compass of Fabricio Mordente. This instrument, which greatly fascinated Giordano Bruno, lay parallel to the table supported by a point beneath its pivot and a pair of points under its extremities. A second pair of points ran on cursors and could be set to divide the arms in any desired proportion. Its superior accuracy arose from the fact that all the points met the paper at right angles. On the reduction compass also it was usual to mark scales for the division of a straight line, or of a circle, into a given number of integral parts. Those scales, however, were not the same as on the proportional compass because of the different relation between pivot and points on the two instruments.

Both the proportional compass and the reduction compass were expensive to make, and both required resetting of their moving parts for nearly every new problem. It appears to have been Guidobaldo del Monte who devised a simpler way of solving mechanically the two basic drafting problems already mentioned. This was the sector illustrated by G. P. Gallucci in a book about mathematical instruments published at Venice in 1598, with a license dated in May 1595, and it was of such recent invention that Gallucci, instead of saying who the inventor was, said only that he was too well known to need mention by name. Muzio Oddi, in 1633, identified this as Guidobaldo's. Again, it was not a calculating sector; its virtue was that there were no moving parts except the pivot, whence it was inexpensive to make and speedy in operation. The points at the ends of the arms were set to the given line or to the diameter of the given circle. To divide a given line into n equal parts, the required segment was the separation of the two points marked n on one face of the sector. To construct a given regular n-gon, the side was given by the separation of the two points marked n on the other face.

Of these drafting instruments only Guidobaldo's sector had any direct connection with Galileo's "compass," and even in that case the connection was more or less accidental. They have been described mainly to enable readers to understand the various attempts that have been made to associate Galileo's compass with drafting instruments of the period.

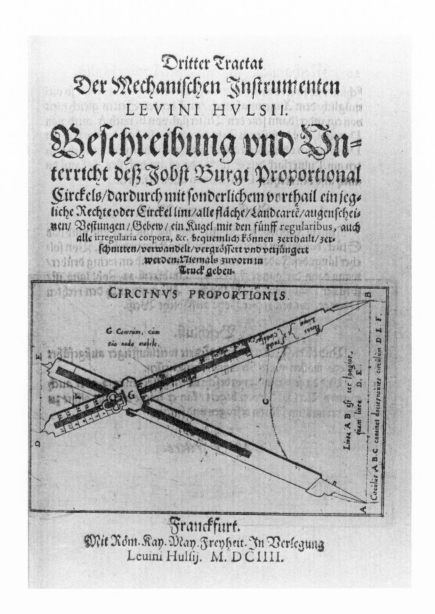

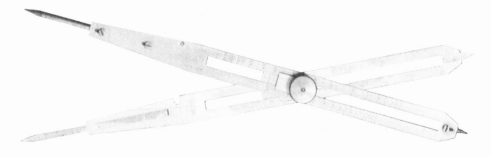

Title page of the book by Hulsius (1604), showing proportional compass credited by
him to Joost Burgi. For comparison, a modern Swiss instrument of this type is shown.
From the author's collection, Toronto, Canada.

13

Fabricio Mordente's reduction compass, at the Adler Planetarium, Chicago, Illinois.

Sector of the Guidobaldo type with added measuring scales along edges. Made at Crema in 1639, showing that these two simple scales, eliminated on Galileo's instruments by 1599, remained long in demand among less-sophisticated users.
From the author's collection.

14

III.

In Galileo's course on military architecture and fortification, he mentioned an artillery elevation gauge that he called the *squadra*. This was in use all over Europe, having been introduced in 1537 by Niccolò Tartaglia of Brescia in a book called "The New Science" devoted to artillery matters. The elevation gauge consisted of a sort of carpenters' square having one leg longer than the other, a plumb line hung from the corner, and a quadrant arc graduated into twelve equal parts, called "points." The longer leg was placed in the mouth of the cannon and held along the lower side of the barrel; the plumb line then showed the "points of elevation" of the gun, point-blank meaning dead level.

Another instrument introduced by Tartaglia in the same book was for use in determining the distance and height of a target by means of sighting and triangulation. It was a square having equal arms with sights along one arm and a smaller square facing the opposite direction; the arms (or *ombre*) of this smaller square were graduated into twelve parts each. Thus the plumb line hung from the corner of the large square must cut one *ombra* or the other, depending on the angle of sighting. Rules were given for triangulation by this instrument, as for several others in use by Galileo's time. One of those was the *archimetro*, probably invented by Ostilio Ricci, Galileo's teacher of mathematics, who wrote a treatise on its uses. This instrument had hinged arms, so that it could be folded shut for greater convenience of carrying on field trips.

In 1595–96 Galileo combined Tartaglia's two instruments into one, making the arms of equal length and providing graduations along a single quadrant arc that would serve the purposes of both Tartaglia's instruments (and some further purposes as well). The quadrant was also made removable, and the arms were hinged so that they could be folded together for carrying. For this instrument, without markings along the arms, Galileo wrote his treatise on triangulation in 1596.

Now, since Galileo's squadra with quadrant removed had the same shape as Guidobaldo's sector, and because surveyors and military architects often wished to divide a line or a circle into equal parts, Galileo engraved those two scales along the inner sides of the arms of his squadra in the form used by Guidobaldo. These two scales were omitted from Galileo's later sectors, but we know where they were on the first model from his 1597 manual of instructions for use of the "military compass." By the time he composed that manual, he had already added several other useful scales, as shown on the schematic diagram.

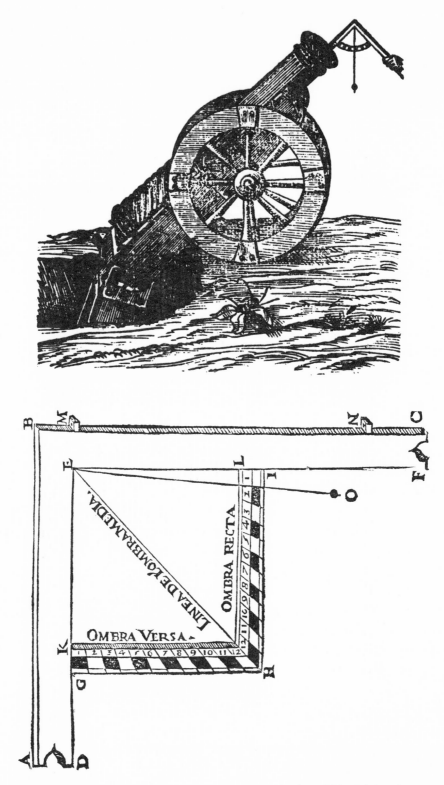

Tartaglia's artillery elevation gauge at 45° or six points (top) and his sighting instrument. From Niccolo Tartaglia's *La Nova Scientia*, Venice, 1537, in the author's collection.

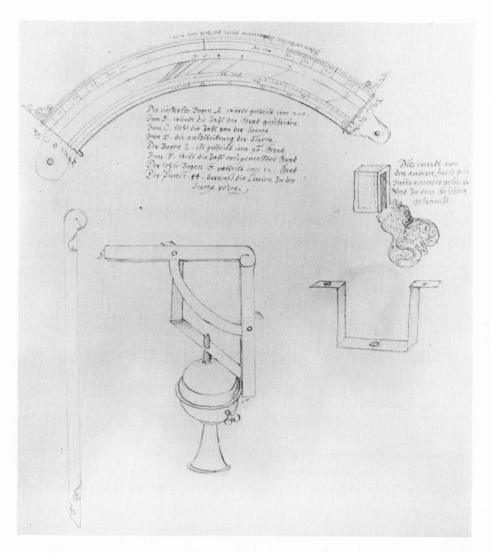

Surveying and gunnery accessories to Galileo's squadra, as drawn by or for a German student of Galileo's at Padua, showing "movable foot," surveying bracket, and universal joint. From an unpublished anonymous contemporary manuscript in the author's collection.

Galileo's Tetragonic Lines permitted the immediate construction of a circle or of any regular polygon (up to 13 sides) having an area equal to that of a given circle, or regular polygon. This gave a quick mechanical solution of "squaring the circle" and also permitted the reduction to square measure of various regular polygons. In fact, it made possible the determination of the area of any figure bounded by straight lines, as Galileo explained in his book.

The Polygraphic Lines gave a direct construction for any regular polygon up to 15 sides, having as one side a given line of fixed length. This was in military practice a more useful scale than Guidobaldo's, because fortifications

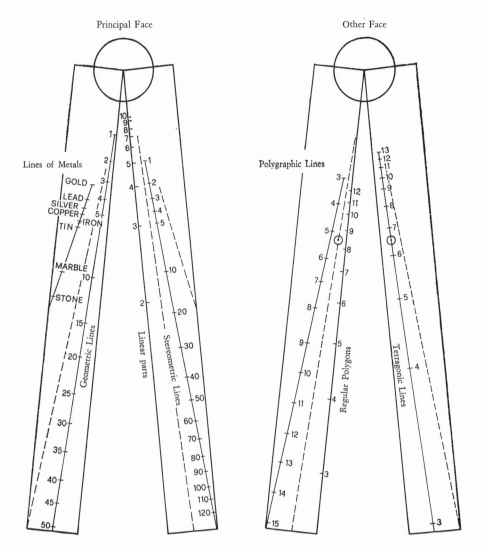

Galileo's first set of scales for the sector represented schematically. Scales for Linear Parts and Regular Polygons, taken from Guidobaldo's sector, were eliminated on the final model. A scale of Equal Parts (Hood's Scale) replaced the former, and the Polygraphic Lines replaced the latter, while Galileo's Added Lines took the positions of the Polygraphic Lines here. Reconstructed by the author from Galileo's early instructions in Volume II, *Edizione nazionale delle Opere di Galileo Galilei*, Antonio Favaro, ed., Florence, 1932.

often used parts of regular polygons and because the length of one side was not infrequently determined arbitrarily by an existing part of an old fort or by some feature of the terrain.

On the main face of Galileo's 1597 sector were two scales designed for mechanical calculation, unlike those which served only to facilitate geometrical constructions. The problem that inspired Galileo to invent them was one that

18

was very difficult to solve by calculations at the time, involving as it did a cubic relationship and also the relative densities of different solids. Galileo called this problem "making the caliber," and it was a highly practical one. Speed and accuracy in solving it might save lives; time and materials were always saved, and skill in solving this problem could mean the difference between military victory and defeat. You will find Galileo's treatment of it in Operation 24 of his book, but from 1597 to 1599 it occupied the first position in his manual of instructions.

The "problem of caliber" was this: Given a charge of powder known to be correct for a certain artillery shot using a cannon of known bore and a cannonball of standard material, to find the equivalent charge for a cannon of any other bore and a ball of any other standard material. Before Galileo tackled the problem it was solved approximately by means of empirical tables that were neither exhaustive nor very reliable when extrapolated to data not specifically covered. The data tabulated were usually carried on sticks rather than on pages of paper; such devices, of very limited utility and accuracy, were described in books published during the English Civil War forty years after Galileo taught his mechanical solution using the sector. An artillery man with no previous mathematical training could learn to use the two scales mentioned to solve any "caliber" problem in a half minute, whereas detailed numerical calculations might take ten minutes and required mathematical skill possessed by very few artillerymen or even officers.

Following the two scales needed for the "problem of caliber" (one of relative densities and the other of volume increments), Galileo placed a scale of area increments corresponding to changes in side, diameter, or radius. Had the next position on this face of his sector not been occupied already by the scale for division of a line into equal parts, Galileo would probably have put there in 1597, by analogy to the preceding volume- and area-increment scales, a simple scale of very small equal divisions of a line. Had he done so, he would have had at once a general-purpose calculating device valuable not only to artillerymen but also to surveyors, mechanics, bankers, merchants, and everyone who in his daily activities deals with problems of proportionality. The equal-divisions scale was in fact the only one on the principal face of Hood's sector of 1598, devised mainly for the use of surveyors.

Galileo's 1597 sector could solve all area problems of figures bounded by straight lines and of circles and semicircles. To attain complete generality in the Euclidean geometrical tradition, all that remained was to find means to deal mechanically with figures bounded by straight lines and circular arcs in any number and any arrangement. This problem occupied Galileo during 1598, and he solved it toward the end of that year or early in 1599 by means of the "Added Lines" described in Operation 32. These lines required a double scale, marked off along both sides of the same lines, one set of numbers being for the half-chord of a circular segment and the other set for the altitude of a segment. For legibility this double scale was best placed near the outer edges of the arms, where on the 1597 model Galileo had put his "polygraphic scale." He now moved this to the inner edges, abandoning Guidobaldo's scale for regular polygons that had previously been placed there. It was no longer needed, because the construction of regular polygons could be equally well carried out by using the Polygraphic Lines, with only one additional and simple operation, as Galileo explained in his new instructions.

Finally, on the other face of the sector, the old scale for equal division of a line into a given number of segments was replaced by the Arithmetical Lines. These solved the original problem with only one additional simple step, and at the same time greatly expanded the range of divisions that could be made. You will see, in Operation 1, how Galileo described procedures which would determine equal divisions more precisely than we ordinarily do with a fine ruler. But the greatest advantage of all was that the Arithmetic Lines, replacing the scale of equal parts, enabled the user of the sector to solve mechanically any rule-of-three problem with sufficient accuracy for practical purposes and with great speed. Since proportionality problems arise continually in every occupation that uses mathematics, it was this scale that ushered in a generally useful mechanical computer. This should be called "Hood's Scale," though without implying that Galileo had first heard of the English device before he could proceed with his own. Nor did he get his idea from any earlier Italian instrument, since none making use of this simple and powerful technique was included in Gallucci's book in 1598. Galileo's Arithmetic Lines were the natural development of the Stereometric and the Geometric Lines which had preceded them, as shown by his original instructions for his first model of which no example is known to have survived.

In or shortly before March 1599, Galileo rewrote his instruction manual to include the new scales and their uses; at that time he eliminated the discussions of the Guidobaldo scales and added sections on map-scaling, extraction of square and cube roots, applications of proportionality problems, and the like. There is a copy of this revision in the Rocco-Bauer collection at the California Institute of Technology, which I described in 1960. It followed the previous order, with the "problem of caliber" placed first, although that problem no longer occupied the most space because of the large field that was opened up by mechanical solution of proportionality problems.

Again the treatise on triangulation was included as an appendix or supplement to the instructions for the calculating sector, but this time a number of additions were made to it. Galileo had always included numerical examples, worked out to illustrate various uses of his "squadra" in surveying. These

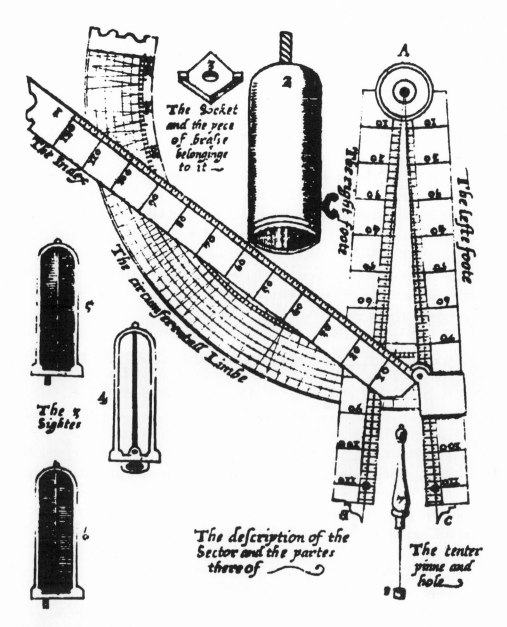

Principal face of Thomas Hood's sector and surveying accessories. Two scales to be placed on the reverse face were described in the text but not illustrated; they gave the sides of regular polygons and fractional areas. From Thomas Hood's *The Making and Use of the Geometrical Instrument Called a SECTOR*, London, 1598, copy in the British Museum Library, London.

examples were still left in, but now he added to them instructions for their solution by using the new Arithmetical Lines. Each of these additions was preceded by words to this effect: "But for those who cannot easily manage numbers, the same result can be found on the Instrument as follows. . . ." This shows that when Galileo wrote his treatise on triangulation in 1596, he did not yet know of any instrument capable of solving proportionality problems mechanically. If such an instrument had been known in Italy, it would have been in use at the Venice Arsenal, for it would have been very useful in shipbuilding and was invaluable in working from models of any kind, as will be seen.

The simultaneous introduction of the basic sector scale in England and Italy is of interest because of the very different backgrounds and purposes of the two men responsible. Thomas Hood was not a professor of mathematics in a university, but a practical mathematician intent on the education of men engaged in useful occupations. In his presentation of the sector he referred not to Euclid, but to Pierre Ramus, who was not of the Sorbonne but of the Collège de France and in his time had been an enthusiast for popular education. In England Hood's instrument remained for a generation confined to solution of relatively simple problems of arithmetic and of lines and areas before it was greatly altered by Edmund Gunter.

Galileo, on the other hand, was a professor of mathematics, concerned at first with a problem that was amenable to systematic treatment only through algebra. In 1597 forty years were yet to elapse before René Descartes applied algebra systematically to geometry, and its application to physics came even later. Hence Galileo first sought a mechanical approximation to a practical problem that still lay beyond the scope of university mathematics. Now, as to simple proportionality problems, there was no reason for Galileo to seek a mechanical aid in their solution, which was easy for him and for his students. It was only in perfecting his sector that he perceived the utility of mechanical calculation to people outside the universities, and it was to the activities of such practical men that his interest began increasingly to turn at Padua around 1600, as he lost interest in the traditional philosophical physics he had been taught at Pisa and began to envision a mathematical physics rooted in mechanics.

IV.

In July 1599 Galileo employed an instrument maker named Marcantonio Mazzoleni to move into his own house, bringing his family, to manufacture for sale the new instrument and such others as were desired by himself or his students. Galileo provided rooms and meals for him and his family, paid

Mazzoleni a small annual salary, and in addition supplied the brass blanks and other materials for which, when engraved and finished, he paid Mazzoleni about two-thirds the price Galileo charged for them. The manufacturing operation was not very profitable, but Galileo made a good deal of money from tuition fees charged to private students who wanted his course on the use of the instrument. A Galilean brass sector cost about 35 lire; a course in its use, about 120 lire. Hence it seems most unlikely that Galileo's sector was a mere elaboration of a well-known and widely used previous instrument. At any rate the young Polish and German gentlemen who paid so generously for his instruction seem not to have known of this supposed earlier instrument, and we shall see in due course what happened soon after Galileo published his book about the uses of his own device.

In or around October 1599 Galileo completely rewrote his instruction manual. This time he placed first the use of the Arithmetic Lines, described common non-military applications of proportionality, and then went on to the other scales in order of complexity, the "problem of caliber" now being moved to Chapter 20. This new instruction manual was provided in manuscript form to owners and buyers of Galileo's sector, and its wording remained unchanged until 1606, when it was finally printed, in the form here translated, with various refinements and last-minute additions.

By 1603 more than forty of Galileo's sectors had been made, according to his statement, and at least twenty had been carried abroad, those being identifiable from his account books. They were usually of brass, although a few had been made of silver, one of which went to the Archduke Ferdinand of Austria (a fact of interest to our story later). Now, during the summer of 1603 Galileo was ill, and one of his German students, about to leave Padua, came to tell him that a Dutchman called Jan Eutel Zieckmesser was showing an instrument much like Galileo's that he claimed to have invented. Some men at Padua, hostile to Galileo, were saying that he had stolen the idea.

As soon as Galileo regained his health, he arranged for a confrontation with the foreigner at the home of G. A. Cornaro, an old and distinguished citizen, in the presence of others. The two devices were seen to be much alike in some respects and quite different in others. Zieckmesser granted that Galileo could not have taken anything from his instrument, since he had never allowed anyone else to examine it before. Galileo, on the other hand, was persuaded that at least some things on Zieckmesser's instrument had been taken from his own. Galileo would indeed have had grounds for that belief if, for instance, his scales for the "problem of caliber" or his "Added Lines" for quadrature of circular segments were present, since those are not likely to have been independently invented by others.

It should be remembered that while many of Galileo's sectors were in the hands of foreigners, together with instructions for their use, Galileo's name did not appear either on the instrument or on the instructions. He did not put his name on the syllabuses he wrote on military architecture, or mechanics, or cosmography either. At times copies were made of these by a professional scribe paid by Galileo, who sold them to students if they wanted them, and in every case where one of these manuscripts survives with a title and the author's name, they were written in later, by the owner, probably after Galileo became famous in 1610. So as things stood, Zieckmesser could have taken one of the Galilean scales from an instrument abroad, or even from manuscript instructions, without knowing whose invention it was. In fact he had been a student at Padua in 1601, apparently without knowing anything about the sector then, since he would hardly have displayed it in Padua as his own invention if he had known it to be Galileo's.

Shortly after this incident, it appears, Aurelio Capra and his son Baldessar asked Cornaro to induce Galileo to teach them the uses of his instrument. (Other evidence points to 1602, which seems less likely.) Aurelio Capra was a fencing master who had brought his young son to Padua from Milan a few years before to enroll him as a medical student. A German astronomer, Simon Mayr, who came to Padua late in 1601 was tutor to Baldessar in astronomy and mathematics. Galileo explained his "compass" to the Capras at Cornaro's house, and also gave or sold one of the instruments to Cornaro, together with the manual of instructions.

In the spring of 1604 Galileo befriended the impoverished Aurelio by recommending him to the Duke of Mantua in connection with some medical secret sought by the duke. This was supposed to have been known to a man called Gromo, leader of an alchemical group at Padua with which the Capras and Mayr were associated. Gromo had died in 1603. Galileo's aid to Aurelio in 1604 was ill repaid; early in 1605 the Capras again approached Cornaro, this time asking him to lend them his Galilean sector and the instructions so that Baldessar might make one for himself. Cornaro complied, and in due course got the material back; what use had been made of it will presently appear.

In mid-1605 Mayr returned to Germany. Galileo spent that summer as special tutor in mathematics to the young Prince Cosimo de' Medici. He was at this time not satisfied with his situation at Padua; having negotiated inconclusively for the post of court mathematician with the Duke of Mantua in 1604, he tried at this juncture for such a position at Florence. As a part of that project he promised to publish his instructions for the "compass" and dedicate the book to Cosimo. This was done during the summer of 1606, only sixty copies being printed and the work being done at Galileo's house. The reason was that Galileo wished to retain his monopoly on the sector, and for the same reason the book included neither a drawing of the instrument nor an explanation of the construction of its scales. The book was intended only for buyers of Galileo's compass, not for potential rivals.

About the end of March 1607 a book appeared over the name of Baldessar Capra that was in many respects a Latin paraphrase of Galileo's book, with an illustration of the instrument and descriptions of the construction of its scales. In the prefatory matter Baldessar implied that the invention was his own, stolen by others who discussed it. A copy of this book was given to Cornaro early in April by Aurelio Capra. The next day Cornaro returned it with a strong letter of protest, and a few days later he supplied Galileo with an affidavit stating that the "compass" dated back to 1597. At that time Baldessar had been only seventeen years old and he had not begun the study of mathematics until 1602. Galileo obtained similar affidavits from others and filed a complaint against Capra with the governors of the university, at Venice. Hearings were held at which Galileo cross-examined the plagiarist, showing him to be ignorant of matters of which he claimed to have been the author and to be unable to explain errors in translation and printing of his supposed book that Galileo himself could account for. The evidence of fraud was clear. Capra was expelled from the university, and the book was confiscated.

By that time, however, some thirty copies of the Capra book had been sent outside the boundaries of Venetian jurisdiction, and Galileo was concerned about his reputation with foreign mathematicians. Even more offensive to him was Capra's implication that in dedicating his own book to Cosimo de' Medici Galileo had presented something that was not his to give. He therefore proceeded to publish an account of the entire matter, in the course of which he had occasion to mention Simon Mayr, though without implicating him in the plagiarism. Years later, after Mayr claimed priority over Galileo in the discovery of Jupiter's satellites (publishing four years after Galileo on that topic), Galileo asserted that Mayr had been responsible for the 1607 plagiarism, although, having departed for Germany before it was published, he could not be prosecuted.

There was probably some truth in this, since among the things proved by Galileo at the Venice hearings in 1607 was the fact that parts of the Latin version published by Capra had been taken not from Galileo's published book of 1606 but from the earlier version circulated in manuscript. It is likely that Mayr, as a mathematician, made Latin translations from Cornaro's manuscript copy before returning to Germany in mid-1605 and that those were incorporated by Capra in the 1607 book. Capra's ignorance at the hearings concerning mathematical aspects of the instrument can hardly be accounted for in any other way. He alone, however, was responsible for publication of the book, whether or not Mayr had written parts of it two years before and had left these with Capra.

V.

In 1610 a German mathematical practitioner, Johann Faulhaber, published a book about a calculating sector nearly identical with Galileo's. In the opening section he said that his acquaintance with it dated from a visit paid to him (probably in 1603) by Mathias Bernegger en route from Austria to Strasburg. Faulhaber had recognized the value of the instrument, although he considered some of its scales less useful than others that he put in their places. He said further that before publishing he had made careful inquiries to determine the name of the original inventor and had learned that this was Galileo Galilei, professor of mathematics at Padua. Because Bernegger seems never to have visited Italy, it is probable that he had seen the silver example sent by Galileo to the Archduke of Austria and in that way knew of its inventor. Faulhaber's inquiries were probably made because of other claimants who had appeared in the meanwhile, possibly including Georg Galgemaier, to be mentioned below.

In 1613 Bernegger published at Strasburg a complete and correct (though unauthorized) Latin translation of Galileo's 1606 book, to which was added the first accurate depiction of the instrument and a clear and correct account of the manner of constructing it. Bernegger's commentaries on the text were of such importance that they were later translated into Italian and published with Galileo's collected works, as well as separately. It was Bernegger who also translated Galileo's famous *Dialogue* into Latin in 1635, after it had been suppressed by Church edict in 1633.

In 1615 Georg Brentl published at Ulm a book on the proportional compass and on the sector "as improved and augmented by Georg Galgemaier." The title page seems to imply that Galgemaier may have previously published on the subject, but I have been unable to find any such book. What Brentl's book contains is virtually a reprint of the book by Hulsius published in 1604, concerning the proportional compass, and a German paraphrase of Galileo's 1606 book (made probably from Bernegger's 1613 Latin translation). No mention whatever was made of Hulsius, Galileo, Faulhaber, or any other earlier writer on these subjects.

In 1616 a rudimentary sector made at Paris (and there called *compas de proportion*) was claimed by D[enis?] Henrion in his mathematical memoirs, according to his own later statement, and its description and uses were published separately by him in 1618. Later editions of this book will be mentioned below.

In 1623 the sector was again taken up in England after a quarter-century of

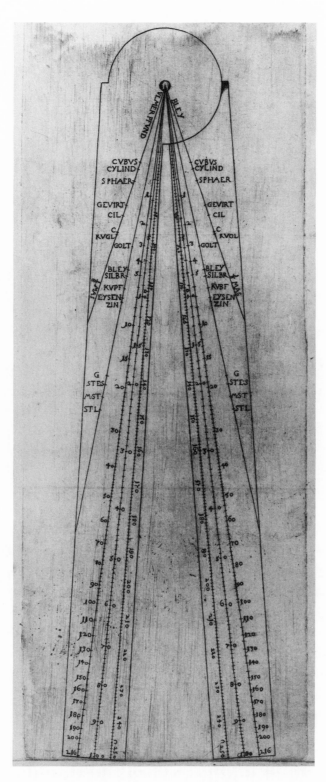

Modified Galilean sector as shown in Johann Faulhaber, *Newe geometrische . . .
Inventiones*, Frankfurt, 1610. From a copy in the author's collection.

27

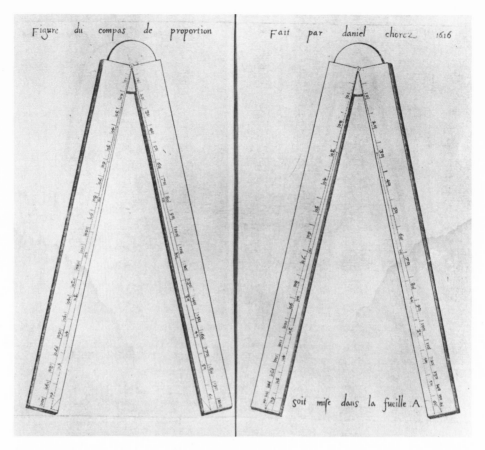

Henrion's primitive French sector of 1616. From D. Henrion's *L'Usage du compas de proportion*, Paris, 1624, (first published in 1618). From a copy in the author's collection.

neglect. This was done by Edmund Gunter, who mentioned neither Hood nor Galileo, and who elaborated the instrument with trigonometric and navigation scales. He was aware of the Galilean scales, saying that others had used scales of little importance which he would put in space otherwise wasted on his sector. Gunter's book went through many editions during the next half-century, and was supplemented by English mathematical editors (Samuel Foster and William Leybourn), but no edition I have seen names any writer other than Gunter as having written earlier on the sector.

Henrion brought out in 1624 a new edition of his book on the so-called *compas de proportion*, mentioned above. In this and all later editions he reproached Gunter (without naming him) for having stolen the instrument— not from Hood or from Galileo but from Henrion. This instance of extreme provincialism as late as 1624 is hard to match in the history of the sector. It is probable that Henrion saw a simple sector made by Daniel Chorez at Paris in 1616 and did not know that anything like it existed outside France until he had seen Gunter's book. Since that had mentioned nothing earlier—not even the pioneer work of Gunter's countryman Thomas Hood—and since Henrion had seen a sector seven years before Gunter's book appeared, he seems to have felt safe in claiming the invention. His book was often reprinted; a very late edition, containing material added to the "fourth edition" of 1631, was printed at Rouen with the mistaken date 1564 on the title page, probably a transposition of 1654 but perhaps an error for 1664. A modern French work on mathematical instruments has attempted to fix the invention in the year of Galileo's birth, and place it in France, using this mistaken title page as the basis for that argument. In 1644, however, Pierre Herigon had already rejected Henrion's claim and had identified Galileo as the original inventor of the sector.

Among all the claimants to the calculating sector or its equivalents, only one (apart from Hood and Galileo) appears to me to deserve attention. This was Michael Coignet of Antwerp, and the instrument was his "pantometric rule" devised in the 1580s. This originally consisted of a brass plate on which were engraved various scales, including some that later appeared on both Galileo's and Gunter's sectors. They were used in conjunction with a pair of ordinary dividers, by means of which trigonometric and other data could be transferred to drawings on paper, which in turn eliminated the need for numerical calculations. This was hardly a mechanical calculating device in the ordinary sense, nor could it have been of use to laymen without mathematical education. Nevertheless it might have led to the invention of the calculating sector. Coignet was in correspondence with Guidobaldo in the 1580s and could have told him of the "pantometric rule"; he also wrote to Galileo in 1588 to compliment him on some theorems on centers of gravity sent to Coignet by Abraham Ortelius at Rome, where Galileo had left the theorems with Christopher Clavius, the leading Jesuit mathematician. I consider it unlikely that Coignet imparted his pantometric rule to Guidobaldo, since nothing of the kind was mentioned in the latter's notebooks or correspondence, and it seems to me virtually impossible that he communicated information about it to Galileo directly. The only extant letter of his to Galileo is one sent in 1588, nine years before Galileo began work on his sector—and then he did not begin with the Arithmetic Lines as one would expect him to have done if he had started from a previously developed calculating ruler.

In 1610 Coignet transferred the scales of his pantometric rule to a pair of sectors (there being too many to accommodate them all on a single sector, even using both faces), and at that time he wrote out instructions for their use. These instructions, dated and left in manuscript, were eventually published in 1626 after Coignet's death. At that time the editor at Paris mistakenly claimed invention of the sector for Coignet, placing this supposed event in the 1580's, an evident confusion with the pantometric rule.

A late English sector of the Gunter type, machine engraved in the nineteenth century.
From the author's collection.

30

VI.

Having studied a good many seventeenth-century sectors made in various places, I am inclined to believe that all those which have exactly the scales described by Galileo, in the same order of arrangement and with the same abbreviations and markings, were made for Galileo either by Marcantonio Mazzoleni at Padua or by makers at Florence, Urbino, and in Germany who were authorized to manufacture them by Galileo himself. The places named were specifically mentioned by Galileo in his defense against Capra in 1607.

Some sectors of the period are very much like Galileo's but have one or more scales omitted, moved, or replaced by others. These were certainly not made for or authorized by Galileo, nor are they to be considered copies of his sector. They were intended either to usurp the invention, as by Galgemaier, or to improve its utility, as by Faulhaber, since scales for the "problem of caliber" were of use only to artillerymen and took up space that could be used for special-purpose scales needed, say, by barrel-makers.

Old sectors frequently exhibit characteristics that are of assistance in identifying their probable countries of manufacture. Among these are patterns of lettering and numbering, the languages abbreviated along the Metallic Lines, kinds of ornamentation, and the scales themselves. Italian instruments seem on the whole to be closest to the Galilean model. English sectors after Hood's depart farthest from that and favor trigonometric scales, also often found on Belgian, Dutch, and French sectors and seldom on Italian or German examples that I have seen. The "meridional line" and "rhumb line" of Gunter's sector seem to be found only on English models. French makers often included a scale of equal volumes for the five regular solids and the sphere, analogous to Galileo's Tetragonic Lines for areas. Special fittings to adapt the sector for direct use in surveying are most often found on Italian instruments, probably because few sectors with a quadrant arc were made outside Italy. That accessory originated with Galileo for military and with Hood for surveying purposes; after Gunter's embellishments the English sector was little used in surveying, if at all, but became a calculating device almost exclusively.

Three examples of the Galilean sector as made at Padua for Galileo by Mazzoleni are known: Galileo's own instrument, now at the History of Science Museum in Florence; an identical instrument but lacking the clinometer lines on the quadrant, preserved at the Castello Sforzesco in Milan; and one in the David Wheatland collection at Harvard University which I have not been able to inspect for possible variants. In my own collection is a sector believed by me

Galilean sector in the author's collection, described in text.

to be one of those authorized by Galileo to be made at Florence, from which the original quadrant is missing and has been replaced by a modern replica. The movable foot is likewise missing, wear along the gilding of the arms showing it to have been much used long ago. This sector is of heavier gauge and is longer than the Paduan model, measuring 285 mm. rather than 250 mm. from pivot to base line. The "Added Lines" are marked with a reversed D, as are all examples, and also with a small square, lacking on the Paduan instruments but prescribed in Galileo's manuscript instruction manuals before 1606. The Florentine lily is used for ornamentation on the screw-heads and at the center of the ornamental scrollwork.

An exact photographic reproduction of the instrument preserved at Florence will be found in the second volume of the National Edition of Galileo's works, edited by Professor Antonio Favaro. The first edition of Galileo's book to carry an illustration of his sector was printed at Padua in 1640, (reproduced on page 38); the engraved illustration was again included in the 1649 edition by the same printer. Among Galileo's papers at the Central National Library in Florence (vol. 36, f. 41) there is a fine drawing of the sector attributed to Vincenzio Viviani, who lived with Galileo from 1639 to his death early in 1642 and who later edited the first collected edition of his works, printed at Bologna in 1655–56, in which there is also an engraved plate showing the Galilean sector.

The translation presented here was made from the National Edition, Volume II, printed in Florence in 1932 by Tipografia G. Barbera.

SELECTED HISTORICAL
BIBLIOGRAPHY ON THE SECTOR

The ensuing bibliography comprises most of the books relevant to the history of the sector during the first century of its existence and the principal secondary sources of more recent times. Books relating only to the proportional and reduction compasses are not listed.

ANCIENT WORKS

Niccolò Tartaglia, *La Nova Scientia* . . . (Venice, 1537)

Thomas Hood, *The Making and Use of the Geometricall Instrument, Called a SECTOR* . . . (London, 1598)

G. P. Gallucci, *Della Fabrica et uso di diversi strumenti* . . . (Venice, 1598)

Galileo Galilei, *Le operazioni del Compasso Geometrico et Militare* (Padua, 1606; 2d ed. Padua 1640; 3d ed. Padua 1649)

Baldessar Capra, *Usus et Fabrica Circini cuiusdam Proportionis* . . . (Padua, 1607)

Galileo Galilei, *Difesa . . . contro alle Calunnie & imposture di Baldessar Capra* . . . (Venice, 1607)

Johann Faulhaber, *Mathematici Tractatus* . . . (Frankfort, 1610), and German language edition: *Newe geometrische und perspectivische Inventiones* . . . (Frankfurt, 1610)

Galileo Galilei, tr. Mathias Bernegger, *De proportionum instrumento . . . ex italica in latinam linguam nunc primum traslatus* . . . (Strasburg, 1613); reissued with new title page in 1635. An edition dated 1612 has been mistakenly reported.

Georg Brentel, *Hern Georgii Galgemairs . . . zubreitung und gebrauch der hochnutzlichen . . . porportional* [sic] *Schregmäss und Circkels* . . . (Ulm, 1615). The title page implies an earlier publication by Galgemaier which I have been unable to find.

D. Henrion, *L'usage du compas de proportion* (Paris, 1618) The first appearance seems to have been in the author's *Mémoires Mathématiques* (1616), followed by the above separate printing and other editions at Paris in 1624 and 1631 (called on title page the fourth edition. An edition at Rouen erroneously dated 1564 must have been printed in 1654 (or 1664?), since it contains not only the entire contents of the 1631 edition but thirty additional pages of demonstrations.

Edmund Gunter, *The Sector* (London, 1623); reprinted 1624, 1636; "fourth edition" 1653(?) and 1662; "fifth edition" 1673.

Michael Coignet, tr. P. G. S., ed. Charles Hulpeau, *La Géométrie réduite en une facile et briève practique, par ... le Pantomètre, ou Compas de Proportion* ... (Paris, 1626). In fact Coignet's original "pantometer" was not described, having been confused by the editor with the 1610 pair of sectors to which its scales were transferred.

Muzio Oddi, *Fabrica et uso del Compasso Polimetro* (Milan, 1633) Oddi considered the calculating sector a simple development from Guidobaldo's sector, noting in the margin later claimants whom he thought to have no standing: Coignet, Galgemaier, Galileo, and Henrion (in that order).

Paolo Casati, *Fabrica et uso del compasso di proportione* (Bologna, 1664); reprinted 1671, 1673, 1685.

Domenico Lusvergh, *Di Galileo Galilei il compasso geometrico adulto per Giacomo Lusvergh* ... (Rome, 1698) The author was an outstanding maker of mathematical instruments, including the Galilean sector in reduced size and without quadrant.

MODERN WORKS

Antonio Favaro, ed., *Edizione Nazionale delle Opere di Galileo Galilei*, Tipografia G. Barbera (Florence, 1890–1909); reprinted 1929–39. "Per la storia del Compasso di Proporzione," *Atti del Reale Ist. Veneto ...*, LXVII (1907–08), pt. 2, pp. 723–39.

Silvio Bedini, "The Instruments of Galileo Galilei" in *Galileo Man of Science*, ed. E. McMullin (New York, 1967), pp. 262–68.

Paul L. Rose, "The Origins of the Proportional Compass from Mordente to Galileo," *Physis* 10 (1968, pp. 53–69).

S. Drake, "An Unrecorded Manuscript Copy of Galileo's *Use of the Compass*," *Physis* 2 (1960), pp. 281–90. "Galileo and the First Mechanical Computing Device," *Scientific American* 234:4 (1976), pp. 104-113. "Tartaglia's Squadra and Galileo's Compasso," *Annali dell'Ist. e Museo di Storia della Scienza di Firenze* 2, (1977), pp. 35–54.

Ivo Schneider, "Der Proportionzirkel, ein universelles Analogrecherinstrument der Vergangenheit," *Deutsches Museum Abhandlungen und Berichte* 38 (1970) 2, pp. 1–96.

LE OPERAZIONI
DEL COMPASSO
GEOMETRICO,
ET MILITARE.

DI
GALILEO GALILEI
NOBIL FIORENTINO

LETTOR DELLE MATEMATICHE
nello Studio di Padoua.

Dedicato

AL SERENISS. PRINCIPE DI TOSCANA
D. COSIMO MEDICI.

IN PADOVA,

In Cafa dell'Autore, Per Pietro Marinelli.   MDCVI.

Con licenza de i Superiori.

OPERATIONS OF
THE GEOMETRIC AND
MILITARY COMPASS

OF

GALILEO GALILEI

FLORENTINE PATRICIAN AND
TEACHER OF MATHEMATICS
in the University of Padua.

Dedicated to
THE MOST SERENE PRINCE
DON COSIMO DE' MEDICI.

PRINTED AT PADUA

in the Author's House, By Pietro Marinelli. 1606

With license of the Superiors.

Translated by Stillman Drake   Florence 1977

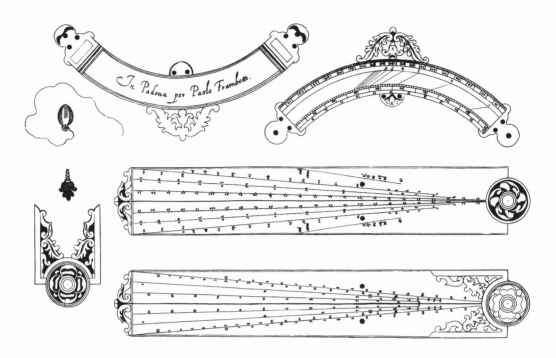

Galileo's Geometrical and Military Compass. This plate, engraved for Paolo Frambotti, was added to the second edition of the *Compasso*, Padua, 1640. A copy was tipped into the Library of Congress copy of the first edition, 1606, which was issued without illustration. From the Bern Dibner collection.

TO THE MOST SERENE
DON COSIMO DE' MEDICI

PRINCE OF TUSCANY, ETC.

f, Most Serene Prince, I wished to set forth in this place all the praises due to your Highness' own merits and those of your distinguished family, I should be committed to such a lengthy discourse that this preface would far outrun the rest of the text, whence I shall refrain from even attempting that task, uncertain that I could finish half of it, let alone all. Besides, it was not to magnify your splendor (which already shines like a rising sun on all the West) that I took occasion to dedicate the present work to you, but rather that this should always carry the embellishment and ornamentation of your name written in front of it, as in my mind, bringing grace and splendor to its dark shadows. Nor do I step forth as an orator to exalt your Highness' glory, but as your most devoted servitor and humble subject I offer you due tribute as I should have done before, had not your tender age persuaded me to await these years more suited to such studies.

I doubt not that this little gift will be gladly received by you, not only because your infinite native gentility so persuades me, but because I am sure this reading is proportioned to your many other regal exercises, besides which experience itself confirms me in this, you having deigned during most of last summer to listen benignly to my oral explanation of many uses of this Instrument. May your Highness therefore enjoy this, my mathematical game so to speak, nobly suited to your first youthful studies. And advancing with age in these truly royal disciplines, expect from my simple mind from time to time those more mature fruits that Divine Grace has conceded it to me to gather. And so, with all humility, I bow to kiss reverently your robe and pray God for your great happiness.   From Padua, the 10th of July 1606.

From your Highness'

Most Humble and Obliged Servant

Galileo Galilei.

TO THE DISCREET READERS

he opportunity of dealing with many great gentlemen in this most noble University of Padua, introducing them to the mathematical sciences, has by long experience taught me that not entirely improper was the request of that royal pupil who sought from Archimedes, as his teacher of geometry, an easier and more open road that would lead him to its possession; for even in our age very few can patiently travel the steep and thorny paths along which one first must pass before acquiring the precious fruits of this science; frightened by the long rough road, and not seeing or being able to imagine how those dark and unfamiliar paths can lead them to the desired goal, they falter halfway there and abandon the undertaking. I have seen this happen the more frequently, the greater the personages I have dealt with, as men who being occupied and distracted by many other affairs cannot exercise in this that assiduous patience that would be required of them. Hence I excuse them together with that young King of Syracuse, and desiring that they should not remain deprived of knowledge so necessary to noble gentlemen by reason of the length and difficulty of ordinary roads. I fell to trying to open this truly royal road—for with the aid of my Compass I do that in a few days, teaching everything derived from geometry and arithmetic for civil and military use that is ordinarily received only by very long studies. I shall not say what I have accomplished by this work of mine, but let those judge who have learned from me up to now, or will learn in the future, and especially those who have seen instruments invented by others for similar purposes—although most of the inventions, and the best that are included in my Instrument, have not previously been attempted (or imagined) by them. Among these, the foremost is that of enabling anyone to resolve instantly the most difficult arithmetical operations; of which, however, I shall describe only those that occur most frequently in civil and military affairs. I regret only, Gentle Readers, that although I have taken pains to explain the ensuing things with all possible clarity and facility, yet to those who must draw them from writing even these will remain cloaked in some obscurity, losing for many people that grace which arouses marvel in seeing them actually performed, and in learning them by word of mouth. But these are matters that do not permit themselves to be described with ease and clarity unless one has first heard them orally and has seen them in the act of being carried out. This indeed would have been a powerful reason for me to refrain from printing this work, had it not come to my ears that another into whose hands my Instrument and its explanation, had come, I know not in what form, was preparing to appropriate it to himself.[1] This made it necessary to insure by printed evidence my labors and reputation against any who wanted to claim it. By way of warning there are not lacking testimonies of princes and other great gentlemen who in the past eight years have seen this Instrument and learned from me its use, of whom it will suffice to name only four. One was the illustrious and excellent John Frederick, Prince of Holsace and Count of Oldenburg, who in 1598 learned from me the use of the Instrument although it was then not yet brought to perfection.[2] Soon afterward I was honored similarly by the Most Serene Archduke Ferdinand of Austria. The illustrious and most excellent Phillip, Landgrave of Hessia and Count of Nidda, studied the said use here at Padua in 1601, and two years ago the Most Serene Duke of Mantua[3] requested its explanation from me.

It may be added that my silence about the construction of the Instrument, which I shall omit at present for its long and laborious description (and for other reasons) will render this treatise quite useless to anyone whose hands it reaches without the Instrument itself. That is why I have had but sixty copies printed, at my house, to be presented together with an Instrument devised and made with that great care which is necessary, first to my Lord the Most Serene Prince of Tuscany, and then to other gentlemen by whom I know this work of mine to be desired. Finally, it being my intention to explain at present mainly those operations of interest to soldiers, I have judged it good to write in the Tuscan language so that the book, coming into the hands of persons better informed in military matters than in the Latin language, can be understood easily by them.   Live happily.

DIVI-

DIVISION OF A LINE

FIRST OPERATION

Coming to the detailed explanation of the operations of this new Geometric and Military Compass, we shall begin first with the face on which there are marked four pairs of lines with their divisions and scales; among those we shall first speak of the innermost, called the Arithmetic Lines from their division in arithmetical progression; that is, by equal additions which proceed out to the number 250. We shall gather various uses for these Lines; and first:

By means of these Lines we can divide any given straight line into as many equal parts as desired, operating in any of the ways set forth below.

When the given line is of medium size, so that it does not exceed the opening of the Instrument, we take its whole length with an ordinary compass[4] and apply this distance crosswise, opening the Instrument, to some number [and its counterpart on the other arm] on these Arithmetic Lines such that above it, on these same Lines, there is a smaller number contained by the selected number as many times as there are parts into which the given line is to be divided. Then the crosswise distance taken between the points bearing this smaller number will doubtless divide the given line into the required parts. For example:

Wishing to divide the given line into five equal parts, we take two numbers of which the larger is five times the other, such as 100 and 20, and opening the Instrument we adjust it so that the given distance (taken with the ordinary compass) fits crosswise to the points marked 100 and 100. Then, not again moving the Instrument, take the crosswise distance between the points on the same Lines marked 20 and 20; undoubtedly this will be one-fifth of the given line. And in the same order we shall find any other division, noticing that large numbers are to be taken (but not exceeding 250), because then the operation turns out to be easier and more exact.

We shall be able to do the same thing operating a different way, and the order will be this. Wishing to divide the line shown below into 11 parts, we shall take one number eleven times the other, as would be 110 and 10, and setting a compass to the whole distance AB we then fit this crosswise (opening the Instrument) to the points 110. Next, being unable to find the distance between points 10–10 along these same Lines, that [region] being occupied by the large hinge, we instead take the distance between points 100–100, narrowing the points of the compass a bit. Then, fixing one of its legs at point B, we mark with the other the point C; the remaining line AC will then be one-eleventh of the whole line AB, and we likewise fix one leg of the compass at A to mark

point

point E near the other end, making EB thus equal to CA. Next, again narrowing the compass a bit, we take the crosswise distance between points 90–90 and apply this from B to D and from A to F, getting two new lines CD and FE which are also one-eleventh of the whole. Transferring with the same order, in either direction, the distances taken between points 80–80, 70–70, etc., we shall find the other divisions, as clearly seen from the diagram.

Now, when the given line is very short and it is to be divided into many parts, as for example the line AB below which is to be divided into 13 parts, we may proceed by [adapting] this second rule.

Let the line AB be extended faintly out to C, and let there be marked along this [extension BC] some other lines, as many as you please, each equal to AB; in the present example let there be six of them, so that AC is six times greater than AB.[5] It is evident that of the parts of which AB contains 13, all AC will contain 91, wherefore taking all AC with a compass we shall apply this crosswise, opening the Instrument, to points 91–91. Next, narrowing the compass a bit to [fit across] points 90–90, we carry that distance from point C in the direction of A. Marking the point near A, this will give us the 91st part of all CA, which is the 13th part of BA. Then, narrowing the compass bit by bit to 89, 88, 87, etc., we transfer those distances from C towards A, finding and marking the other little parts of the given line AB.

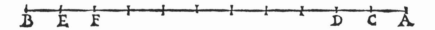

Now finally, if the line to be divided is very long, so that it much exceeds the widest opening of the Instrument, we can nevertheless take in it the required part, which let be (for example) one-seventh. To find this, we first think of two numbers of which one is 7 times the other, as for instance 140 and 20. Open the Instrument at will and take crosswise the distance between points 140–140; see how many times this is included in the given line. However many times it is contained, that many times is the crosswise distance between points 20–20 to be repeated along the large line, and you will have its one-seventh part if the distance taken between points 140–140 precisely measured the given line. If it did not exactly measure this, it will be necessary to take one-seventh of the excess in the manner previously explained, and by adding that to the distance which was laid off many times along the large line, you will have its one-seventh part to a hair, just as was desired.

HOW IN ANY GIVEN LINE WE CAN TAKE AS MANY
parts[6] as may be required.

Operation II.

his operation is the more useful and necessary according as without the help of our Instrument it may be [the more] difficult to find such divisions, which can nevertheless be found instantly with the Instrument. Thus when one is asked to take in a given line some parts, such as (for example) 113 parts out of 197, one just takes the length of the line with a compass and opens the Instrument so that this length fits crosswise between the points marked 197–197; without again moving the Instrument, take with the same compass the distance between points 113–113, and that will doubtless be the fraction of the line requested, equal to one-hundred-thirteen one-hundred-ninety-sevenths of it.

HOW THESE SAME LINES GIVE US TWO, AND
even infinitely many, scales for altering one map into another,
larger or smaller.

Operation III.

t is evident that whenever it is required to draw from a given diagram another similar one, large or smaller in any desired ratio, we must use two scales accurately divided, one of which serves to measure the drawing already made, and the other for marking the lines of the drawing to be made. Two such scales we shall always have from the Lines of which we are now speaking. One will be the line already divided lengthwise along the Instrument; this established scale will serve us for measuring the sides of the given drawing. The other, for drawing the new design, must be alterable; that is, it must be capable of increase or diminution at our pleasure, according as the new drawing is to be larger or smaller. This variable scale will be that which we have crosswise from the same Lines by narrowing or widening the opening of our Instrument. For a clearer understanding of the manner of putting these Lines to use, let us take an example. Thus, let diagram ABCDE be given, to which we are to draw another, similar, but based on line FG which is to be homologous (that is, corresponding) to line AB. Here it is evident that two scales must be used, one to measure the lines of diagram ABCDE and another by which will be measured the lines of the diagram to be drawn, and this latter must be greater or less than the other scale according to the ratio of line

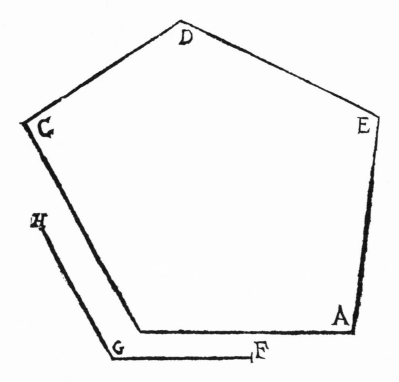

FG to line AB. Therefore take with a compass the length AB, putting one leg of the compass at the pivot of the Instrument and the other at any point it strikes, [establishing a measure for the line] which in our example let be point 60. Then take line FG with the compass, and placing one of its legs at point 60, open the Instrument until the other leg falls crosswise exactly on the other corresponding point 60, without again altering the opening of the Instrument. All the other sides of the given diagram will be measured in turn along the lengthwise scale, and immediately the corresponding crosswise distances to these will be taken for the sides of the new diagram. Thus (for example) we wish to find the length of line GH corresponding to BC; take distance BC with the compass and apply this from the pivot of the Instrument lengthwise along the scale, and holding one leg on the point where it falls, say for example at 66, fit the other leg crosswise to the corresponding point 66, according to which measure you will mark line GH, and it will correspond to BC in the same ratio as that of line FG to AB. Notice that when you want to transform a small diagram into one somewhat larger, it will be necessary to use the two scales in the opposite order—that is, use the lengthwise scale for the diagram to be drawn, and the crosswise scale to measure the lines of the given diagram. Thus, for example, let us take the diagram ABCDEF which we wish to transform into another, somewhat larger, and based on line GH which is to correspond

to

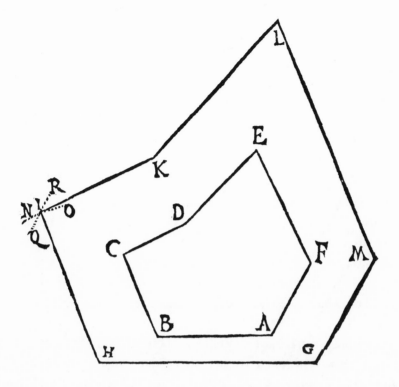

to line AB. To set the scales, take line GH and see how many graduations it contains along the lengthwise scale; seeing this to contain, say, 60, take its corresponding line AB and fit that crosswise to points 60–60, not moving the Instrument again thereafter. Then to find line HI, corresponding to BC, take BC with the compass and find which points it fits along the crosswise scale. Finding this to fit, say, points 46–46, at once take the distance to point 46 lengthwise along the scale and you will have the length of line HI, corresponding to BC. Notice that in this (as in the other) operation it does not suffice to have found the length HI without also finding the point toward which this must be directed so that angle H will be made equal to angle B. Therefore having found [the length of] this line HI, fix one leg of the compass at point H and, with the other, mark faintly a short arc as shown by our dotted line OIN; next the distance between points A and C will be taken, and the number of graduations out to this along the crosswise scale will be sought. This being found to fit, say, at 89–89, take the distance 89 lengthwise with the compass, and holding one leg at G mark the intersection of arc RIQ with the first arc OIN, this being point I toward which you must direct line HI. Then doubtless angle H will equal angle B and line HI will be proportional to BC. In the same way will be found the other points K, L, and M, corresponding to corners D, E, and F.

THE

THE RULE-OF-THREE SOLVED BY MEANS OF A
Compass and these same Arithmetic Lines.

Operation IV.

he present Lines are used not only for resolution of various linear problems, but also for some arithmetical rules, among which we place this one corresponding to what Euclid teaches us: Given three numbers, to find the fourth proportional. This is the Golden Rule, called rule-of-three by practitioners who find the fourth number proportional to the three given. To demonstrate the whole thing by an example, we say for clearer understanding:

If 80 gives us 120, what will 100 give us? Here you have three numbers given in this order: 80 120 100. To find the fourth number which is sought, take lengthwise on the Instrument the second of the given numbers, which is 120, and apply this crosswise to the first, which is 80; then take crosswise the third number, which is 100, and measure that lengthwise along the scale. What you will find—that is, 150—will be the fourth number sought. Notice that the same will happen if instead of taking the second number you take the third, and then in place of the third you take the second; that is, the same will result from the second number taken lengthwise and fitted crosswise to the third and then taking the third crosswise and measuring it lengthwise, as given by the third taken lengthwise and fitted crosswise to the first, thereafter taking the second crosswise and measuring it lengthwise; for in both ways we shall get 150. It is good to notice this, because according to different circumstances one [or the other] way of working will turn out to be the more convenient.

Some cases may occur in the operation of this rule-of-three that could give rise to some difficulty if we are not aware how to proceed in them. And first, it may sometimes happen that of the three given numbers, neither the second nor the third taken lengthwise *can* be fitted crosswise to the first, as when we ask "25 gives me 60; what will 75 give?" Here both 60 and 75 exceed double the first number, which is 25, so that neither one can be taken lengthwise and then applied crosswise to 25. Therefore to carry out our intent we shall take either the second or the third lengthwise and fit it crosswise to *double* the first, that is, to 50; and if double is still not enough, then we shall fit it to triple, or quadruple, etc. Then taking the other one crosswise, we shall affirm that what is shown us by measuring lengthwise the *half* (or third or fourth part) of what was sought. Thus in the example given, 60 being taken lengthwise and fitted crosswise to the double of 25 (that is, to 50), and 75 being then taken crosswise and measured lengthwise, we shall find that it gives us 90, whose double (or 180) is the fourth number that was sought.

Besides this, it may happen that the second or third given number cannot be fitted to the first because the first is so large that it exceeds the largest number marked on the Lines, which is 250; as when we say "280 gives me 130; what will 195 give me?" In such cases let 130 be taken lengthwise and put crosswise to the half of 280, which is 140; next take crosswise half the third number (195), or 97½. That distance measured lengthwise will give us 90½, which is what we sought.

It will be good to add a further warning, to be used when the second or third of the given numbers is very large, the other two being of medium size, as when we say "If 60 gives me 390, what will 45 give me?" Take 45 lengthwise and fit it crosswise to 60, but then, being unable to take 390 whole, we shall take whatever part of it we like. For example I shall take 100 crosswise, which measured lengthwise will give me 75, and since 390 is 90 taken once and 100 taken three times, I shall take the 75 (already found) three times, and add the 67½ that was found for 90; the sum is 292½, the fourth number that was sought.

Finally we may say also how the same rule operates for very small numbers even though on the Instrument we were able to mark only the points from 15 on, because of the hinge that joins and unites the arms of the Instrument. On this occasion we shall use tenths of the graduations as if they were units when saying (for example) "if 10 gives 7, what will 13 give?" Being unable to take 7 and fit it to 10, we shall take 70 (that is, 7 tens) and fit this to 10 tens (that is, to 100). Next taking 13 tens [crosswise], we measure that distance lengthwise and find it to contain 91 graduations, which means 9¹⁄₁₀, we having (as said) made every tenth count for one unit.

From all these instructions, well practiced, you will be able to find easily the solution in every case of any difficulties that may arise.

THE INVERSE RULE-OF-THREE RESOLVED BY
means of the same lines.

Operation V.

y operations not dissimilar are resolved questions of the inverse rule-of-three; here is an example. What [amount of] food that suffices for 60 days to maintain 100 soldiers will sustain as many for 75 days? These numbers arranged by the rule will stand in this order: 60 100 75. Operation on the Instrument requires that you take lengthwise the first number, which is 60, and apply it crosswise to the third number, which is 75; then without moving the Instrument you take crosswise 100, which is the second number, and measure

this

this lengthwise, finding 80, which is the number sought. Here you should notice also that we would find the same by fitting the second lengthwise to the third crosswise and then measuring [lengthwise] the first, taken crosswise. You should also note that all instructions given above for the rule-of-three are again to be exactly followed in this.

RULE FOR MONETARY EXCHANGE.

Operation VI.

y means of these same Arithmetic Lines we can change every kind of currency into any other, in a very easy and speedy way. This is done by first setting the Instrument, taking lengthwise the price in the money we want to exchange, and fitting this crosswise to the price in the money into which exchange is to be made. We shall illustrate this by an example so that everything is clearly understood. For instance we want to change [Florentine] gold *scudi* into Venetian ducats; since the price or value of the ducat is 6 *lire* 4 *soldi*, it is necessary (because the ducat is not an exact multiple of the *lira* and those 4 *soldi* enter in) that we work out both currencies in terms of *soldi*, considering that the *scudo* is priced at 160 *soldi*, the price of the ducat being 124. Hence to set the Instrument for changing gold *scudi* into ducats, take lengthwise the value of the *scudo*, which is 160, and opening the Instrument fit this crosswise to the value of the ducat, which is 124. Then, not moving the Instrument again, whatever amount is given in *scudi* and is to be changed into ducats is taken crosswise and measured lengthwise. For example, we want to know how many ducats make 186 *scudi*; take 186 crosswise and measure this lengthwise. You will find 240, and that many ducats will make 186 *scudi*.

RULE FOR COMPOUND INTEREST, ALSO CALLED
"New Year's gain."

Operation VII.

e can very speedily solve questions of this kind with the aid of the same Arithmetic Lines, and operating in two different ways, as will be made clear and evident by the two following examples. It is asked what will be gained on 140 *scudi* in five years at the rate of 6 percent per annum, leaving the interest on the capital and on the previous interest so that all continually go on earning. To find this, take the initial capital (that is, 140) lengthwise and

fit this crosswise at 100–100. Then without moving the Instrument take crosswise the distance between points 106–106, which is 100 with [a year's] interest, widening the Instrument to fit this distance taken on a compass and applied to 100–100. Next, opening the compass a bit more, take with it crosswise the distance that is now between points 106–106, and again widening the Instrument a little, fit the distance just found to 100–100. Once again open the compass to take this 106–106, and in brief go on repeating the same operation as many times as the number of years of earning; thus in the present example the earnings for five years mean repeating the operation five times. Finally, measuring lengthwise the interval you shall have reached, you will find this to contain 187⅓ graduations, and that is the number of *scudi* which the original 140 have become by compounding at 6 percent for five years. Observe that when it would be more convenient to use 200 and 212 in place of 100 and 106, as often happens, the result will be the same.

The second way of working requires no change in setting of the Instrument after its initial adjustment; it goes as follows, using the same problem as before. To set the Instrument take 100 together with the first [year's] interest, making 106, lengthwise; open the Instrument and fit this crosswise to 100–100, never changing the Instrument again. Then take the amount of money, which was 140, measure this lengthwise, and you will see that the capital and increase after the first year is 148⅖. To find [this for] the second year, take this 148⅖ crosswise and (of course) measure it lengthwise; you will find 157⅓ for the second year. Next take this number 157⅓ crosswise, again measure lengthwise, and you will find 166¾ for capital and earnings at [the end of] the third year. Take this 166¾ crosswise and measure it lengthwise, and for the fourth year you will have 176¾; finally, take this crosswise and again measure lengthwise, and for the fifth year you will have 187⅓ as capital and earnings. For more years, if you wish them, go on repeating the operation. Notice that if the original capital is a sum that exceeds 250, [the largest] marked on the Arithmetic Lines, you must operate by parts, taking a half, a third, a quarter, a fifth, or any other part of the given sum; at the end, by taking two, three, four, five, (or more) times what you have found, you will know the desired amount.

OF THE GEOMETRIC LINES

WHICH FOLLOW NEXT, AND THE USES
thereof. And first how by means of these we can increase or decrease
in any ratio all areas of figures.

Operation VIII.

he lines that next follow the Arithmetic (explained above) are called the Geometric Lines from their being divided in geometrical progression out to 50. From these we gather various uses; and first, they serve us for finding the side of a plane figure that has a given ratio to another [similar] that is given. For example, given the triangle ABC, we wish to find the side of another triangle that has the [area] ratio to it of 3:2. Select two numbers in the given ratio; let these be for instance 12 and 8. Taking line BC with a compass and opening the Instrument, fit this to points 8-8 of the Geometric Lines; then, without changing the opening, take the distance between points 12–12. If we now make a line of that length the side of a triangle, corresponding to line BC, the surface will doubtless be three-halves that of triangle ABC.

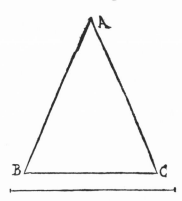

The same is to be understood of any other kind of figure, and likewise we can do the same with circles, making use of their diameters or radii as of the sides of rectilineal figures. And let people with little schooling note that the present operation shows the increase or diminution of all plane surfaces; as for example having a map that contains, say, 10 *campi* of land,[7] we want to draw one that contains 34. Take any line of the ten-*campi* map and apply this crosswise to points 10–10 of the Geometric Lines; then, without again moving the Instrument, take the crosswise distance between points 34–34 of the same Lines; upon such a length describe your map similarly to the first, according to the rule taught above in Operation 3, and you will have the desired map containing exactly 34 *campi*.

HOW WITH THE SAME LINES WE CAN FIND THE
ratio of two similar plane figures.

Operation IX.

Let there be given, for example, two squares, A and B, or indeed any two other figures of which those two lines A and B designate homologous sides. We want to find what ratio there is between the areas. Take line B with a compass and, opening the Instrument, fit this to any pair of points of the Geometric Lines, say to points 20–20. Then, not altering the Instrument, take line A with the compass and see what number this fits when applied to the Geometric lines; finding it to fit, say, at number 10, you may say that the ratio of the two areas is that which 20 has to 10; that is, double. And if the length of this line does not fit exactly at any of the graduations, we must repeat the operation and, trying other points than 20–20, get both lines exactly fitted at some [marked] points, when consequently we shall know the ratio of the two given figures, that being always the same as that of the two numbers of the two points at which the said lines fit for the same opening of the Instrument. And given the area of one of two maps you will find the area of the other in this same way, as for example: The map with line B being 30 *campi*, how large is map A? Fit line B crosswise to points 30–30 and then see to what number line A fits crosswise; that many *campi* you shall say are contained in the map with line A.

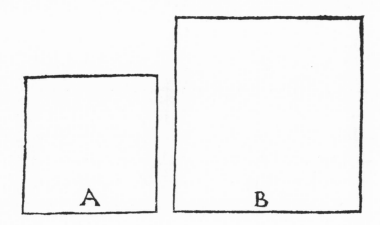

HOW YOU CAN CONSTRUCT A PLANE FIGURE
similar and equal to several others given.

Operation X.

et there be given for example three similar figures of which lines A, B, and C are homologous sides; we are to find a single figure similar to these and equal to all three. Take with a compass the length of line C, and opening the Instrument fit this to whatever number along the Geometric Lines you wish. Say it was applied to points 12–12; then, leaving the Instrument in that position, take line B and see what number it fits on the same Lines. Let this be for example at 9–9, and since the other fitted at 12–12, add together the two numbers 9 and 12, keeping in mind 21. Then take the third line, A, and in the same way see what number on the same Lines it fits crosswise; and finding it to fit, say, at 6–6, add 6 to the 21 you were keeping in mind, for 27 in all. Then take the crosswise distance between points 27–27 and you will have line D, upon which you construct a figure similar to the other three that were given, and this will be of a magnitude equal to all those three together. In the same way we may reduce to one any number [of figures] that are given, provided that they are all similar to one another.

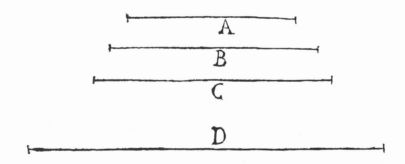

GIVEN TWO UNEQUAL SIMILAR FIGURES, TO
find a third, similar to these and equal to their difference.

Operation XI.

he present operation is the converse of that already explained in the preceding chapter,[8] and it is carried out in this way. Given, for example, two unequal circles of which the diameter of the greater is line AA and that of the smaller is line BB, to find the radius of the circle that is equal [in area]

to

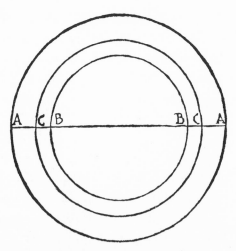

to the difference between circles A and B. Take with a compass the length of line A (the larger), and opening the Instrument apply this length to any point you like in the Geometric Lines; let it be fitted, for example, to the number 20–20. Without altering the Instrument, see at what point of the same Lines you can fit line B. Finding this to fit, say, at number 8–8, subtract this from 20 and get 12. Taking the distance between points 12–12 you have line C, [the diameter] whose circle will equal the difference between the two circles A and B. What is thus exemplified in circles by way of their radii[9] is to be understood to be the same when we operate with one of their homologous sides.

SQUARE ROOT EXTRACTION WITH THE HELP
of these same lines.

Operation XII.

n the present chapter three different ways of proceeding in the extraction of square root will be explained: one for numbers of medium size, one for large numbers, and the third for small numbers, meaning by "numbers of medium size" those in the region of 5,000, by "large" those around 50,000, and by "small" those around 100. We shall begin with medium-sized numbers first.

To find and extract the square root of a given medium number, then, the Instrument must first be set. This is done by fitting crosswise to 16–16 on the Geometric Lines, the distance of 40 graduations taken lengthwise along the Arithmetic Lines. Then take away from the given number its last two digits, which denote the units and tens, the number thus left being taken crosswise on the Geometric Lines and measured lengthwise along the Arithmetic; what is found will be the square root of the given number. For example, you wish to find the square root of 4,630. Take away the last two digits (the 30) and 46 remains; therefore take 46 crosswise on the Geometric Lines and measure this lengthwise along the Arithmetic. There you will find it to contain 68 graduations, which is the approximate square root sought.

Two things are, however, to be noted in using this rule. The first is that when

the

the last pair of digits (taken away) exceeds 50, you should add a unit to the number that remains. Thus if, for instance, you want to take the root of 4,192, then since 92 exceeds 50 you should use 42 instead of the 41 that remained; for the rest, follow the above rule.

The other caution to be noted is that when what remains after removing the last two digits is itself greater than 50, then since the Geometric Lines do not go beyond 50 you must take the half, or some other [aliquot] part of the remaining number, and using this distance you must geometrically double or multiply the number [obtained] according to the part taken;[10] the final distance, thus multiplied, when measured lengthwise along the Arithmetic Lines, will give you the root you sought. For example we want the root of 8,412. The Instrument having been set as above, and the last two digits being removed, there remains 84, a number not on the Geometric Lines. So you take its half, or 42, and then having taken crosswise the distance between points 42–42, you will have to double this [distance] geometrically. This can be done by widening the Instrument until the said distance fits to some number of the Geometric Lines for which a double exists [on those Lines], as for example would be done by fitting it to 20–20 and then taking the distance between points 40–40. This, measured finally [lengthwise] along the Arithmetic Lines will show you approximately 91⅔, which is about the root of the given number 8,412. And if you had been obliged by the given number to take one-third, to triple that geometrically you fit it crosswise to a number of the Geometric Lines for which there is a triple, as it would be for 10 to take 30, or for 12 to take 36.

As to the manner of dealing with large numbers, that differs from the above only in the setting of the Instrument and in your taking away the last *three* digits from the given number. To set the Instrument take 100 lengthwise along the Arithmetic Lines and fit this crosswise to points 10–10 on the Geometric. This done, to get the square root of 32,140, for example, take away the last three digits, leaving 32, and take this crosswise on the Geometric Lines; measuring this then along the Arithmetic, you have 179, the approximate root of 32,140. The same caution noted in the preceding operation should be strictly observed in this, for when the three digits taken away exceed 500 you must add a unit to the remaining number, and if the latter exceeds 50 you must take a part of it (half, one-third, etc.), geometrically doubling or tripling what you get for the part taken, in the manner explained above.

For small numbers set the Instrument in the first way, putting 40 against 16–16, and then take the given number crosswise on the Geometric Lines, without having taken away *any* digits. Measuring this distance lengthwise along the Arithmetic Lines[11] you will find the desired root as a whole number

with

with fraction. Note that here the tens of the Arithmetic Lines must serve you as units, and the units as tenths of a unit. For example, we want the root of 30. Set the Instrument as said by placing 40 taken lengthwise along the Arithmetic Lines at 16–16 on the Geometric, from which take crosswise the distance between points 30–30 and measure this lengthwise along the Arithmetic. You will find 55 graduations, which here means five units and five tenths (that is, 5½), which is the approximate root of 30. Note that in this rule you should again observe the instructions and cautions taught for the other two rules.

RULE FOR ARRANGING ARMIES WITH UNEQUAL
fronts and flanks.

Operation XIII.

o arrange the front line equal to the flank, it obviously suffices to take the square root of the given number of soldiers. But when we wish to arrange an army having a given number of soldiers so that front line and flank are unequal and in a given ratio, then it is necessary to proceed differently in resolving the problem, as explained in the following example.

Let it therefore be given to us to find the front and flank of 4,335 soldiers arranged in such a manner that for every five forming the front there shall be three along the flank. To carry this out with the help of our Instrument, consider first the numbers of the assigned ratio, 5 to 3; add a zero to each, pretending they mean 50 to 30. To find the front, we shall take 50 lengthwise along the Arithmetic Lines, and using a compass, fit this distance crosswise to the Geometric Lines at the number obtained by multiplying together the numbers of the given ratio (which in the present example gives 15); with the Instrument left at that setting, we take crosswise on the Geometric Lines the distance between points marked by a number to be determined as follows. Remove the tens and units from the number of soldiers given, leaving in the present example 43; that distance measured lengthwise along the Arithmetic Lines will give us the front line of such an army, which will be 85 soldiers. The flank will be found in the same way, taking 30 lengthwise along the Arithmetic Lines and setting this crosswise to 15–15 on the Geometric Lines, where we next take crosswise the interval between points 43–43, and this measured lengthwise along the Arithmetic Lines will give us 51 for the flank. The same procedure will hold for any other number of soldiers and for whatever ratio is required of us, noting (as was said of square root) that when the units and tens taken away from the given number exceed 50, a unit is added to the hundreds that

57   remain

remain, etc. Nor do I wish to omit that having found the front line by the rule explained above, you can find the flank by another and speedier means and using only the Arithmetic Lines, working as follows. Having already found 85 for the front in the above example, and the numbers of the ratio being 3 to 5, which is as if we said 50 to 30, or 100 to 60; etc., the 85 taken lengthwise along the Arithmetic Lines can be fitted crosswise to 100–100 on those same Lines, and at once the interval crosswise can be taken between points 60–60 on them; this, measured lengthwise, will show us the same number, 51, that was found by the other manner of working.

And this operation explained under the example of armies is to be understood as the rule for one class of algebraic equations—the one for the square of an unknown set equal to a number,[12] whence all the questions solved by that [algebraic rule] are also resolved by operating with our Instrument in the way just explained.

DISCOVERY OF THE MEAN PROPORTIONAL BY
means of these same lines.

Operation XIV.

ith the help of these Lines and their divisions we can with great ease find between two lines, or two numbers, the mean proportional line or number, in this manner. Let the two given numbers, or the two measured lines, be for example 36 and 16; take with a compass the length of one, say 36, and opening the Instrument fit that to points 36–36 of the Geometric Lines; then without moving the Instrument take the distance between points 16–16 of the same Lines and measure this along the same scale, finding it to be

24 graduations, which is precisely the mean proportional number between 36 and 16. Note that to measure the given lines we may make use not only of the scale marked on the Instrument, but of any other whatever when that of our Instrument is too small for our purposes.

Note furthermore that when the lines and the numbers that measure them are very large (that is, exceed 50 which is the largest number marked on our Geometric Lines), our intent can nevertheless be carried out by operating with

[aliquot]

[aliquot] parts of the given numbers, or with other smaller numbers having the same ratio as the original numbers; and the rule is this. Say that we want to find the mean proportional number between 144 and 81, both of which exceed 50. Take 144 lengthwise along the Arithmetic Lines, to be applied crosswise on the Geometric Lines; since there is no number in the latter that large, I shall in imagination take a part of this number 144, which shall be

(say) one-third or 48, and then fit the distance so taken crosswise to

points 48–48 of the Geometric Lines. Then, thinking

of one-third of 81 (the other given number),

which is 27, I shall take that number

crosswise on the same

Geometric

Lines, and this, measured along the Arithmetic, will

give me the mean proportional

sought, which is

108.

OF THE STEREOMETRIC LINES,

AND FIRST, HOW BY MEANS OF THESE ALL
similar solids can be increased or diminished in any given ratio.

Operation XV.

he present Stereometric Lines are so called because their divisions are according to the ratios of solid bodies, out to 148; from them we shall collect many uses. The first of these will be that proposed above; that is, given one side of any solid body how we may find the side of another [similar one] that has a given [volume] ratio to it. For example, let line A be the diameter of a sphere, or to speak more familiarly a ball, or the side of a cube or other solid, and let it be required to find the diameter or side of another similar solid having to it the ratio which 20 has to 36. Take line A with the compass, and opening the Instrument fit that to points 36–36 of the Stereometric Lines; this done, take next the distance between points 20–20, which will give line B, the diameter (or side) of the solid which is to the other, whose side is A, in the given ratio of 20 to 36.

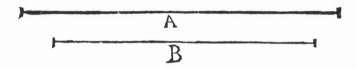

GIVEN TWO SIMILAR SOLIDS, TO FIND THE
Ratio between them.

Operation XVI.

he present operation is not much different from the one explained above, and can be solved with great ease. If therefore, we are given the two lines A and B, [sides of similar solids], and are asked what ratio exists between their similar solids, we shall take one of these with the compass, for example A, which we shall apply (opening the Instrument) to some number on the Stereometric Lines. Let it fit, say, at 50-50; taking next the length of the other line, B, see at what number it can be fitted, and it being found (for example) to fit at 21–21, we shall say that solid A has to solid B the ratio of 50 to 21.

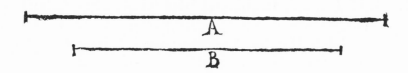

GIVEN ANY NUMBER OF SIMILAR SOLIDS, TO
find a single one equal to all of them together.

Operation XVII.

iven the three lines A, B, and C, sides of three similar so-lids, we want to find one equal to all these. To do this, take with a compass line A, which is applied to some point on the Stereometric Lines; let this be for example to points 30–30. Not moving the Instrument, consider the number at which line B fits; and finding it to fit, for example, at 12, add that number to the number 30 already mentioned, getting 42, to be held in mind. Then taking line C with the compass, consider the num-ber along the Stereometric Lines at which this fits, and let this be at 6–6, for example. Adding this number to the other, 42, we shall have 48, so that taking the distance between points 48–48 the line D will be found, whose [similar] solid will equal all three given ones, A, B, and C.

EXTRACTION OF CUBE ROOT.

Operation XVIII.

e shall explain two different ways for the investigation of the cube root of any number.

The first will serve us for medium numbers, and the second for large ones, meaning by "medium numbers" those not ex-ceeding 148 after we have removed its units, tens, and hundreds.

To extract the cube root of these, the Instrument will first be set by fitting crosswise to points 64–64 of the Stereometric Lines the distance 40 taken lengthwise along the Arithmetic Lines. That done, remove the final three digits from the given number and take what remains crosswise on the Stereometric Lines; and measuring this then lengthwise on the Arithmetic Lines, what is found will be the cube root of the given number. For example we seek the cube root of 80,216; the Instrument having been set as above, and the final three digits being removed, 80 remains; therefore take 80 crosswise on the Stereometric Lines and measure this lengthwise along the Arithmetic. You will find 43, which is the approximate cube root of the given number. Note that if, when the last three digits are removed, more than 148 remains, that being the largest number on the Stereometric Lines, you must then operate by parts. For example, seek the cube root of 185,840. Upon removing the final three digits (840), 186 remains—I say 186 although 185 was left, because here the hundreds digit removed was greater than 5, meaning more than half a thousand, whence to seek a better round thousand I add one more unit so that 186 remains—and since this exceeds 148, let us take its half, which is 93, crosswise on the Stereometric Lines (already set). This distance must then be stereometrically doubled,[13] so fit it crosswise to some number of the Stereometric Lines which has a double [thereon], and the latter being taken crosswise and then measured along the Arithmetic scale will be the cube root sought. Staying with the above example, let us fit the distance between points 93–93 (already taken) to points 40–40 on the Stereometric Lines; then take the distance between 80–80 and measure this along the Arithmetic. This shows us 57, which is the approximate cube root of the given number.

The other way of operating (for large numbers) will be to set the Instrument by fitting the distance of 100 graduations taken lengthwise along the Arithmetic Lines to 100–100 crosswise on the Stereometric. Let it be so set; then from the given number remove the last *four* digits and take the remaining number crosswise on the Stereometric Lines, measuring this then lengthwise along the Arithmetic. For example, the number 1,404,988 being proposed, and the Instrument having been set in the above way, remove the final *four* digits, leaving 140. Take that number crosswise on the Stereometric Lines and measure this lengthwise along the Arithmetic, giving us 112, the approximate cube root of the given number. Do not forget that when the remaining digits exceed 148, the largest number on our Lines, one must operate by parts, as in the other rule explained above.

Operation XIX.

hen there shall be proposed to us two numbers or two measured lines between which we are to find two others that are the mean proportionals, we shall be able to do this easily by means of the present Lines as will be clear from the following example. Given the two lines A and D, of which one is for example 108 and the other is 32, take the larger with a compass and fit it, opening the Instrument, to the numbers 108–108; then take the distance between points 32–32, which will give the length of the second line, B; and measuring this by the same scale with which the given lines were measured, 72 will be found. Then to find the third line, C, fit anew points 108-108 on the same Stereometric Lines, this time to the length of B, once again finding the distance between points 32–32; this will now be the magnitude of the third line, C, and measured on the same scale as before, it will be found to give 48 graduations. Notice that it is not necessary to take first the larger rather than the smaller line, but operating with either in the same way you will always find the same [result].

HOW EVERY PARALLELOPIPED CAN BE RE-
duced to a Cube by means of the Stereometric lines.

Operation XX.

et there be given the parallelopiped solid whose dimensions are unequal, as 72, 32, and 84; what is sought is the side of the cube equal to it. Take the mean proportional between 72 and 32 in the way explained in Operation 14—that is, take 72 lengthwise along the Arithmetic scale and set this crosswise

against

against 72–72 on the Geometric Lines (although because they do not go that far, you set it against one-half, or 36, and then take next the other number, that is, 32–32, crosswise on the same Lines, or rather you take its half, or 16–16, having likewise cut the original 72 in half); what is found will clearly be the mean proportional number between 72 and 32. Then measure this along the Arithmetic Lines and you will find it to be 48, whence you now set this crosswise to the same number, 48–48, of the Stereometric Lines, and without altering the Instrument you take crosswise the third number of the given solid, which was 84. The operation will then be completed, because by making this line [the distance between points 84–84] the side of a cube, that will truly be equal to the given solid. Measuring it along the Arithmetic scale you will find it to be about 57½.

EXPLANATION
OF THE METALLIC LINES,
MARKED NEXT TO THE STEREOMETRIC.

Operation XXI.

he present Lines have divisions to which are affixed these symbols: *Au, Pb, Ag, Cu, Fe, Sn, Mar, Sto,* which mean Gold, Lead, Silver, Copper, Iron, Tin, Marble, and Stone. From these you can get the ratios and differences of [specific] weight found between the materials thus designated, in such a way that with the Instrument set at any opening, the intervals between any correspondingly marked pair of points will give the diameters of balls (or sides of other solid bodies) similar to one another and equal in weight. Thus, whatever be the weight of a ball of gold whose diameter is equal to the distance *Au-Au,* such is also the weight of a lead ball whose diameter is the distance between points *Pb-Pb,* and of a marble ball whose diameter is the distance between points *Mar-Mar* [at the same setting].

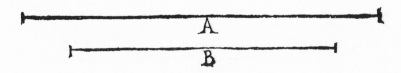

From this we may know at once how large to make a body of any one of the above materials so that it will weigh equally with another, similar in shape but made of another of the said materials—an operation that we shall call "transmutation of material." For instance, if line A is the diameter of a tin ball and we want to find the diameter of a gold ball equalling it in weight, we take with the compass line A, and opening the Instrument we fit that to the points *Sn-Sn;* then we immediately take the distance between points *Au-Au,* and that will be the diameter of the gold ball (represented by line B) which equals in weight the other, of tin. And the same is to be understood of all other solid bodies of the other designated materials. Now if we combine the use of these Lines with that of the previous Lines, we shall collect therefrom many important advantages, as will be explained below. And first:

HOW

HOW WITH THE AFORESAID LINES WE CAN

find the ratio of weight between all metals and other materials marked along
the Metallic Lines.

Operation XXII.

e wish, for example, to find the ratio of [specific] weight
between the two metals silver and gold. With a compass
take the distance between the pivot of the Instrument and the
point marked *Ag*; with the Instrument opened, fit this [cross-
wise] to any number you please along the Stereometric Lines,
say for example to points 100–100. Then, without altering the Instrument, take
the distance between the pivot and the point *Au* and see what number this fits
along the Stereometric Lines; finding, for example, that it fits at points 60–60,
you will say that the ratio of specific weight of gold to silver is as 100 to 60.
Notice that in operation, the diameters taken and fitted to the Stereometric
Lines exhibit inversely the ratio of weight for the metals, so that as seen from
the above example the diameter of the silver gives you the weight of the gold,
while the diameter of the gold gives the weight of the silver. Thus we come to
understand how gold is heavier than silver in the ratio of 40 percent, since 40
is the difference between the two weights found, for gold and silver [gold
being 100].

From this we learn the solution of a very pretty question, which is: Given
any shape of one of the materials marked along the Metallic Lines, to find how
much of some other of the said materials will be needed for forming another
solid equal to it. For instance, we have a marble statue and we want to know
how much silver would go to make one of the same size. To find this weigh
the marble statue, and suppose its weight to be 25 pounds; then take the
distance between the pivot of the Instrument and point *Ag*, the material of the
future statue, and opening the Instrument fit that distance [crosswise] to the
Stereometric Lines at the points marked with the number of the weight of the
statue (that is, to points 25–25); without changing the Instrument you then
take the distance between the pivot and point *Mar*, and see to what number
of the Stereometric Lines it fits crosswise. Finding that it fits at points 96–96,
you will say that 96 pounds of silver are required to make a statue equal in size
to that of the marble one.

COMBINING THE USES OF THE METALLIC AND

the Stereometric Lines, and given two sides for two similar solids made of different materials, to find the ratio of their weights.

Operation XXIII.

ine A is the diameter of a copper ball and B is the diameter of one of iron; we want to know the ratio of their weights. Take line A with a compass, and opening the Instrument fit this to the points of the Metallic Lines marked *Cu-Cu*. Without changing this setting, immediately take the distance between points *Fe-Fe*, which shall be line X. Now, X being unequal to B, and being the diameter of an iron ball equal in weight to A, it is evident that the difference [in diameter] between the two balls A and B will be the same as the difference between X and B. Since X and B are of the same material, their difference is easily found with the Stereometric Lines, as explained before in Operation 16; that is, we shall take line X and (opening the Instrument) fit it to some number, as for instance 30–30, which done we shall see at what point line B fits. Finding it, for example, to fit at 10–10, we shall say that the copper ball [of diameter] A is triple that of the iron ball [of diameter] B.

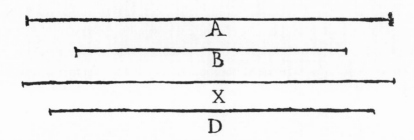

The converse of the foregoing operation can be done equally easily with the same Lines, that is: Given the weight and the diameter (or side) of a ball (or other solid) of one of the materials marked on the Instrument, how one may find the size of another similar solid of some other of the said materials which will weigh equally with any given weight. For example, line X being the diameter [this time] of a marble ball weighing 7 pounds, find the diameter of a lead ball that weighs 20 pounds. Here it is seen that we must perform two operations; one is to transmute marble into lead, while the other is to increase the weight of 7 to 20. The first operation is performed with the Metallic Lines, fitting diameter X crosswise to points *Mar-Mar*, and then, without changing the Instrument, taking the distance between points *Pb-Pb*, which will give the size of the lead solid that weighs as much as the given marble, or 7 pounds.

But

But since we want 20 pounds, we have recourse to the aid of the Stereometric Lines; and fitting this distance to points 7–7 thereon, we at once take the crosswise distance between points 20–20, which will equal line D; and doubtless that will be the side of the solid lead figure that will weigh 20 pounds.

HOW THESE LINES SERVE UNIVERSALLY FOR
the Gunners' calibration for all cannonballs of any material and of every weight.

Operation XXIV.

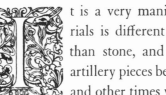It is a very manifest thing that the weight of different materials is different and that iron is [specifically] much heavier than stone, and lead than iron; from which it follows that artillery pieces being fired now with stone balls, again with iron, and other times with lead balls, the same cannon that carries so much [weight] of lead ball carries less [weight] of iron, and still less of stone, whence different charges must be used for the different balls. Therefore those tables for firing (or "calibers") in which are shown the diameters and weights of iron balls cannot be used also for stone balls, but [for the same weight] the diameters must be increased or diminished according to different materials. Moreover it is manifest that the [standard] weights used are different in different countries, and in not only every province, but in almost every city—from which it follows that the calibration suited to the weights of one place cannot be used for the weights of another, and according as the "pound" shall be greater or less in one place than another, it will be necessary for fixed divisions of caliber to carry [shots] longer or shorter distances. From which we may conclude that a caliber adapted to any kind of material and every difference in weight must necessarily be variable and capable of increase or decrease. Such is precisely what is marked on our instrument; for opening this more, or less, the intervals between the divisions found on it increase or diminish without in any way changing their ratios.

Having stated these things generally, let us pass to the particular application of this calibration to any differences in weight and to all the various materials. And since one can come to knowledge of anything unknown only by means of something else, known, it is necessary to know [in our case] only the diameter of a single ball, of any material whatever, and any weight that corresponds to the "pounds" customary in the country where we wish to use our Instrument. From that single diameter we shall, by means of our calibration, come to know the weight of any other ball, of whatever other

material

material—meaning, however, materials marked on our Instrument.

Now the manner of obtaining such knowledge may be made manifest easily by an example. Suppose for instance that we are at Venice, and we wish to make use of our calibration to know the range of some artillery pieces. First we get the diameter and the weight of a ball made of any of the materials designated on our Instrument; suppose for instance that we have the diameter of a 10-pound lead ball in Venetian weight. This diameter we shall mark by two points [scratched] along the side of one arm of the Instrument. Then when we wish to accommodate and adjust our calibration in such a way that by taking the bore of a [Venetian] piece of artillery and carrying this [measurement] to our calibration, we learn how many pounds of lead ball this [gun] carries, we need only take on a compass that diameter of ten pounds of lead (already marked along the side of our Instrument) and then open the Instrument until the said diameter fits to points 10–10 of the Stereometric Lines. Set in that way, those Lines will serve us for exact calibration, so that given the diameter of the mouth of any artillery piece, and transferring it to the said calibration, we shall know (from the number of the points to which it fits) how many pounds of lead ball the said cannon will bear. And if we want to set the Instrument so that the calibration shall correspond to iron balls, then we still take the same diameter of ten pounds of lead marked on the side, and fit this [crosswise] to the points marked *Pb-Pb* on the Metallic Lines; then without changing the Instrument we shall take with a compass the distance between the points marked *Fe-Fe*, which will be the diameter of an iron ball weighing ten pounds. Opening the Instrument, this diameter will be fitted to the points marked 10–10 on the Stereometric Lines, and then said Lines will be exactly set for calibration of iron balls. By similar operations the Instrument can be set for stone balls.

Notice that since it is necessary to mark along the side of the Instrument various diameters of balls corresponding to the [standard] weights of various countries, we should in order to avoid confusion always mark diameters of balls of lead weighing ten pounds, which we shall find to be larger or smaller according to the differences in [local] weights. It will not be difficult to mark down these diameters without the necessity of actually finding ten-pound lead balls, by what was taught before in Operation 23. There, given the diameter of a ball of whatsoever weight and made of any material, we saw how one finds the diameter of any other, of different weight and of any material—meaning always those materials marked on the Metallic Lines. Thus, finding ourselves in any country, provided only that we can find a ball of marble or stone or any other material designated on our Instrument, we can instantly get the diameter of a ten-pound lead ball [in that country's weight].

HOW, GIVEN A BODY OF ANY MATERIAL, WE

can find all the particular measures of one of different material that shall weigh
a given amount.

Operation XXV.

A mong the uses that can be drawn from these same lines, one is that we may increase or diminish solid shapes according to any desired ratio, changing the material or leaving it unchanged, as will be understood from the ensuing example.

We are given a little tin model of a cannon from which we must derive all the particular measurements for a large copper cannon weighing, say, 5,000 pounds. First we shall weigh our little tin model; let its weight be 17 pounds. Next we take one of its measurements, whichever you prefer; let this be for example the thickness of the cannon at its mouth. Opening the Instrument, we apply this to the points *Sn-Sn* of the Metallic Lines, tin being the material of the given model; and since the large cannon is to be made of copper, we at once take the distance between points *Cu-Cu*, which will be the thickness at the mouth of a copper cannon that would weigh the same as the tin one. But since it is to weigh 5,000 pounds, and not 17 like this tin one, we next have recourse to the Stereometric Lines and fit that distance (just found between the points *Cu-Cu*) to the points marked 17–17. Without moving the Instrument we then take the distance between points 100–100, which will give the thickness at the mouth of a [copper] cannon weighing 100 pounds. But we want the weight to be 5,000 pounds, whence this distance must be increased in the ratio of 50:1. Therefore, opening the Instrument farther, we put this [last distance] at some number for which there is another [on the Stereometric Lines] fifty times as great, as would be the case if we fitted it to the points 2–2 and then took the distance between points 100–100, which would doubtless be the measure of the thickness that must be given at the mouth. In this order will be found all the particular measurements for all other parts, such as the throat, the trunnions, the breech, and so on.

No less can we find the length of the cannon, although we cannot open our Instrument to any such distance. To find this, we take on our little model not the entire length, but only a part, such as one-eighth or one-tenth or the like; this may be increased in the way already explained, to show us ultimately one-eighth or one-tenth of the whole length of the large cannon.

But in this [operation] another difficulty [about metals] may arise. Yet our Metallic Lines, designed so as to give measures for transmuting one simple metal into another, can do the same for an alloy of metals when in the above example we must make the cannon not of pure copper, but of metal mixing

copper

copper and tin, as it is the common custom to do. Wherefore, for complete satisfaction, we shall show how, with the aid of those same Metallic Lines, we can find the measurements for any alloy just as for a simple metal. This will be done by adding two tiny points along the Metallic Lines—and I mean very tiny, so that we may remove them after they have served our purpose. Suppose, for example, that the cannon we want to make is to be not of copper (as previously assumed), but must be cast from bronze, in an alloy of three parts copper to one of tin. Then we must carefully divide, on both arms, that short line between the points marked *Cu* and *Sn* into four parts, leaving three of these on the side *Sn* and only one toward *Cu*, and precisely there making our tiny point appear. We then use this point (marked, as was said, on both the Metallic Lines) for our transmutation of metal, just as above we used the points *Cu-Cu*. Using this rule we can, as occasion arises, mark new points for any alloys of two metals in any ratio desired.

It will not be irrelevant and without useful purpose to note, especially when one must make the transmutation into some metal mixed and alloyed of two others in any ratio, that when a single one of the measurements sought has been found by working with great precision in the way explained above, one may on the strength of this unique determined measurement go on to find all the rest by means of the Arithmetic Lines [alone], in a way not much different from that explained in Operation 3. For example, line A was the diameter or thickness of the mouth of the given cannon model, and line B was then found

A ———————— B ————————————————————————————

to be the mouth of the 5,000-pound cannon to be made of a metal containing three parts of copper to two of tin. I say next that to find all the remaining dimensions, we may use the Arithmetic Lines, taking line B and applying it crosswise to any point along those Arithmetic Lines, and the greater the number chosen the better it will be. So let us fit B to the last point, that is, at 250–250, and without moving the Instrument let us see where line A fits crosswise; say this is at 44–44. From this we learn that since A of the model measures 44 graduations, what must correspond to it in the real cannon will be 250 of those same graduations. This same ratio must then hold for every other measurement, whence to find (for example) the thickness of the throat of the real cannon you will take that thickness from the little model and fit it crosswise to points 44–44 of the Arithmetic Lines, taking then (also crosswise) the distance between points 250–250, which will give the throat of the large cannon. By the same rule will be found all the other measurements.

Moreover, to find first with the utmost precision line B, corresponding to the

point

point for the alloy of the two metals assigned,[14] one may proceed thus. First represent separately the two simple metal measures corresponding to tin and copper as the two lines CD and CE, of which CD is the measure corresponding to pure copper and CE to pure tin. Their difference, being line DE, is to be

C   D   F   E

divided according to the assigned ratio of the alloy, whence for three parts copper to two parts tin you will cut line DE at point F so that FE (towards tin) is two parts, and FD (towards copper) is two parts. This is done by dividing all line DE into five parts, leaving three of these towards E and two towards D [by marking F]. Line CF will now be our chief one, as line B was before, in proportion to which, by simply using the Arithmetic Lines in the manner taught in Operation 3, all the other measurements will be found without recourse again to the Metallic and Stereographic Lines.

OF THE POLYGRAPHIC LINES,

AND HOW BY THESE WE CAN DESCRIBE REGU-
lar Polygons; that is, figures of many equal sides and angles.

Operation XXVI.

urning the Instrument over we see the innermost Lines on its other face, called Polygraphic from their principal use, which is to describe on a given line figures having as many equal sides (and angles) as required. This is easily done by taking with a compass the length of the given line and fitting it to the points marked 6–6; next, without changing the Instrument, we take the distance between the points marked with the same number as the number of sides of the figure we wish to draw. For instance, to describe a [regular] figure of 7 sides we take the distance between points 7–7, which will be the radius of the circle containing the heptagon to be drawn. Place one leg of the compass first at one end and then at the other end of the given line and trace with it a short intersection of arcs; then taking this as center we shall describe with the same compass-opening a faint circle. Passing through the ends of the given line, this will contain that line exactly 7 times inside its circumference, whereby the heptagon is drawn.

DIVISION OF A CIRCLE INTO AS MANY PARTS
as is desired.

Operation XXVII.

ith these lines the circumference can be divided into many parts, working by the converse of the preceding operation. Take the radius of the given circle and fit it to the number designating the parts into which the circle is to be divided; then take always the distance between points 6–6 [on the Polygraphic Lines so separated] and this will divide the circumference into the desired [number of] parts.

TETRA-

EXPLANATION
OF THE TETRAGONIC LINES,

AND HOW BY THEIR MEANS ONE MAY SQUARE
the Circle, or any other regular figure, and may also transform all these, one
into another.

Operation XXVIII.

hese Tetragonic Lines are so called from their principal use,
which is to square all regular areas and the circle as well,
which is done by an easy operation. For when we wish to
construct a square equal to a given circle, all we have to do is to
take its radius with a compass, and opening the Instrument fit
this to the two points along the Tetragonic Lines marked by two tiny circles;
then, not moving the Instrument, if you take on the compass the distance
between the points marked 4–4 on those Lines, you will have the side of a
square equal [in area] to the given circle. Likewise if you want the side of the
pentagon, or the hexagon, equal to the same circle, you will take the distance
between points 5–5, or 6–6; for they are respectively the sides of the pentagon
and the hexagon equal to that same circle.

Moreover if we want the converse (given a square or other regular polygon,
to find a circle equal to it), we take one side of the given polygons and fit it
to the points of the Tetragonic Lines corresponding to the number of sides of
the given figure; then, without moving the Instrument, take the distance
between the little circle marks, and by making this the radius you will describe
the circle equal to the given polygon. In conclusion, you can in this way
find the side of any regular figure that is equal to any other.

For example, we wish to construct an octagon equal
to a given pentagon [both regular]. The
Instrument is set so that the side of
the given pentagon fits
at points 5–5,
and
without changing the Instrument the distance
between points 8–8 will give the
side of the octagon
sought.

GIVEN DIFFERENT REGULAR FIGURES, HOW-
ever dissimilar to one another, how to construct a single one equal to all of
them together.

Operation XXIX.

olution of the present problem depends on the preceding one and on Operation 10, explained before. Let there be given to us a circle, a triangle, a pentagon, and a hexagon; we are asked to find one square equal to all the said figures [combined]. First, by the preceding operation we find separately four squares, equal to the given four figures; then, by Operation 10, we find a single square equal to those four, which will doubtless be equal to all four given figures.

HOW ONE MAY MAKE ANY DESIRED REGULAR
figure that equals any given irregular (but rectilineal) figure.

Operation XXX.

he present operation is no less useful than interesting, showing us not just the way to square all irregular [rectilineal] figures, but how to reduce them either to a circle or to any other regular figure you like. Since every rectilineal figure is resolvable into triangles,[15] once we can make a square equal to any triangle we shall make separately individual squares, equal to each of the triangles into which the given rectilineal figure is resolved, and then by Operation 10 we reduce all these squares into a single square, and obviously the square will have been found that is equal to the given rectilineal figure. By means of the Tetragonic Lines we can at will convert that square in turn into a circle, a pentagon, or any other regular rectilineal figure. Thus the solution of the present question is reduced to our having to find a square equal to any given triangle, which you will have very easily from the ensuing lemma.

Operation XXXI.

et it be proposed to make a square equal to the given triangle ABC. Draw to one side two lines at right angles, DE and FG; then take a proportional compass[16] of which one end opens to double the other and hold one of the longer legs at corner A; open the other leg until, when rotated, it just grazes the opposite line BC. Now invert the [proportional] compass and mark with the shorter legs the distance FH, which will be half the perpendicular that falls from angle A on the opposite side BC. This done, take line BC with the longer legs and carry this to FI; holding one leg at point I, move the other out to point H. Again inverting the compass, without further widening or narrowing it, mark with the short legs the distance IK, and holding one of their points at K let the other cut the vertical FG at point L. Thus we now have line LF, which is the side of the square that is equal to triangle ABC.

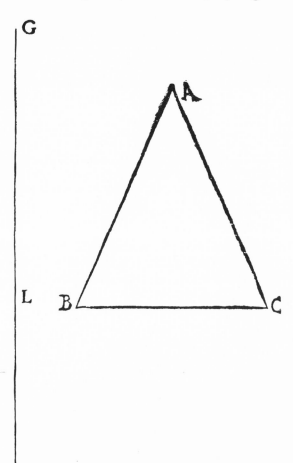

Notice that although we have set forth this operation as performed linearly without the Instrument, that is not because this could not also be easily found on the Instrument. For if we wished to reduce some triangle to a square, as for instance triangle ABC, then taking the vertical that falls from angle A upon the opposite side BC we would see how many graduations this contains

along

along the Arithmetic scale, as say 45, and fit that distance crosswise to points
45–45 on the Geometric Lines; then taking half the line BC we would see
likewise how many graduations along the Arithmetic Lines it contains; finding it to contain, say, 37, we then take crosswise on the Geometric
Lines the distance between points 37–37,
which would give us line LF,
the square on which will
equal triangle
ABC.

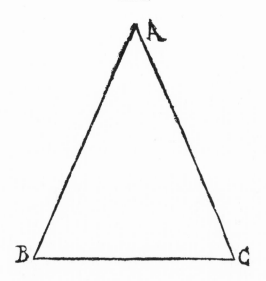

OF THE ADDED LINES, FOR THE QUADRATURE
of segments of a circle and of figures containing parts of circumferences,
or straight and curved lines together.

Operation XXXII.

here remain finally the Added Lines, so called because they
add to the Tetragonic Lines what with those was left to be
desired—that is, a way of squaring segments of circles and the
other figures mentioned in the heading above, which will be
more specifically explained below. These Added Lines are
marked with two series of numbers, of which the outer series begins at this
mark ⊂, followed by the numbers 1, 2, 3, 4, and so on out to 18. The inner
series begins from this mark ⌐⌐,[17] going on then to 1, 2, 3, 4, and so on, also
out to 18. By means of these Lines we can (for the first time) square any given
segment of a circle not more than a semicircle. In order that this may be better
understood, their use will be explained by example.

We

We wish, for instance, to find the square equal to the segment of a circle, ABC. Bisect its chord AC at point D, and take the distance DC with a compass. Opening the Instrument, fit this across the points marked ⊓⊓, and leaving the Instrument at this setting, take next the altitude of the segment (that is, line DB) and see at which points of the outer scale this fits crosswise. Let this be, say, at points 2–2. This done, we must next take with the compass the the distance between points 2–2 of the *inner* scale and form our square on a line of that length, which [square] will be

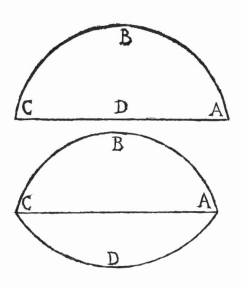

equal to the segment ABC. Now if we have an area contained by two segments of circles with a common chord, as in the next diagram ABCD, we could easily reduce this to a square by drawing the chord AC which divides this figure into two circular segments, whereupon by the rule already given there are found two squares equal to the two separate segments which by recourse to Operation 10 are reduced to a single square, completing the task.

By a not dissimilar operation we can also square any sector of a circle; for drawing the chord of its arc cuts that into one circular segment and one triangle, two parts which can easily be reduced to two squares by the things already taught, and those two then [reduced] to a single square.

Finally it remains for us to show how the same Lines can serve us to square a segment larger than a semicircle, or a trapezium contained by two straight lines and two arcs (like that in the next figure, ABCD), or the lune similar to X; all of which operations have the same resolution. As to the segment larger than a semicircle, if we square the excluded smaller segment in the foregoing manner, and subtract such a square from the square equal to the entire circle, the square equal to the difference will obviously be equal to the larger segment of the circle. Likewise, having found the square equal to the whole segment BAFDC and subtracted from this the square equal to the segment AFD, the remainder will equal the trapezium [ABCD]. And proceeding similarly for the lune X, draw the common chord of the two segments of circles and take separately the squares equal to those two segments; their difference will be the square equal to the lune. As to finding the difference between two given squares and reducing this to another square, that was explained by use of the Geometric Lines in Operation 11.

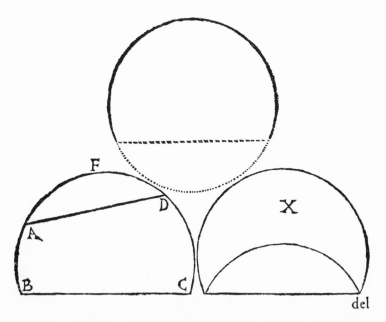

del

ON THE OPERATIONS OF THE QUADRANT.

y attaching the Quadrant Arc to the Instrument we have the gunners' Squadra, divided as is customary into 12 "points."[18] The ordinary use of this is to place one of its legs in the bore of a cannon, having first hung a plumb bob by a thread from the pivot of the Instrument; this thread cutting the circumference, we are shown the elevation of the cannon in "points" as 1, 2, or 3. But since to use the Squadra in this way is not without peril, [the gunner] having to go outside the ramparts or shields and expose himself to the enemy, another way has been thought of to do the same thing in safety—that is, by applying the Squadra near the fuse of the cannon [to its upper surface]. But because the internal bore is not parallel with the outer surface, the metal being thicker near the breech, one must provide against error by lengthening the leg· of the Squadra on the side toward the mouth, adding to this [length] by a movable footing. Thus, placing the Squadra [upright] near the fuse [of a level cannon], we lengthen the forward leg by this footing until the vertical [thread] cuts point 6 [on the usual scale of 12]; and fixing the footing with its set-screw, we mark a little line on the side of the Instrument at the end of the cursor carrying the footing, so that on every occasion of its use [with this cannon] we can place it exactly. Then when we want to fire "at one point" of elevation, the cannon is elevated until the thread cuts the number 7; and if we want "two points" it must cut the 8, and so on.

Next following is the division of the Astronomical Quadrant, whose use (having been treated by many others) will not be explained here.

The circumference that comes next is seen to be divided by some slanted

lines

lines, used for taking the slope of any [rampart] wall, beginning with those which for every ten units of height have one unit of horizontal advance and going on out to those which have one unit of horizontal advance for every 1½ units of height. To employ this [scale of the] Instrument we must hang the [plumb] line from that little hole which is seen at the beginning of the gunners' Squadra;[19] then, approaching the sloping wall, hold the other edge of the Instrument against it and note where the thread cuts [this scale]; if it cuts at number 5, for example, we shall say that the wall has one *braccio*[20] of horizontal advance for every 5 *braccia* of height; or, cutting at number 4, that it has one unit of horizontal advance for every 4 of height.[21]

VARIOUS WAYS OF MEASUREMENT BY SIGHT-
ing; and first, of Vertical Heights whose bases we can approach, and [from which we can] retire.

he final circumference, divided into 200 parts, is a scale for measuring heights, distances, and depths by means of sighting. And beginning first with heights, we shall show various ways of measuring them, starting with vertical heights whose bases we can reach, as it would be if we wished to measure the height of the tower AB. Being at point B, let us move toward C and walk 100 paces or other units. Stopping at C, let us sight along one side of the Instrument

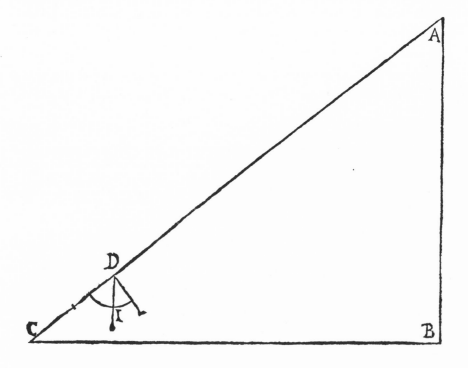

at

at the height A, as is seen for side CDA, and note the graduations cut by the thread DI. If these fall in the 100 away from the eye, as in the given example for arc I,[22] the number of those graduations is the number of paces (or other units we have measured along the ground) that we shall say are contained in height AB. But if the thread cuts the other 100, as seen in the next diagram in which we want to measure the height GH, our eye being at I and the thread cutting points M and O, then take that number of graduations and divide it into 10,000; the result will be the number of units contained in height GH. For example if the thread cuts point 50 [in the arc near the eye], then divide 50 into 10,000 and get 200, and that many units [of which 100 equal IH] will be contained in height GH.

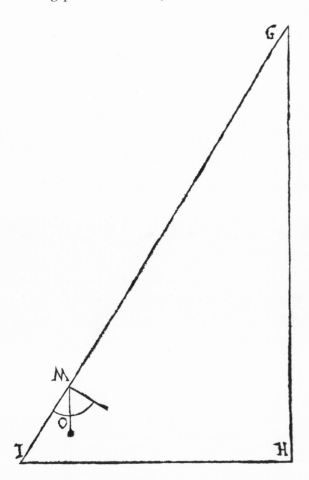

We have seen that sometimes the thread will cut the 100 away from the side along which we sight, and sometimes it will cut the 100 touching that side; either may happen in many of the ensuing operations. Therefore as a general rule it is always to be remembered that when the thread cuts the first 100, contiguous to the sighting side, one must divide 10,000 by the number cut by the thread, following in the rest of any operation whatever rule is written there; for in the ensuing examples we shall always assume that the thread cuts the second 100 [away from the eye].

The better to make clear the multitude of uses of this Instrument of ours, I wish to simplify the more laborious calculations required by the rules for measurement by sighting; they can be performed without trouble and very swiftly by using a compass along the Arithmetic Lines.[23] But starting from the above operation, for those who cannot divide 10,000 by a number cut by the vertical [thread], I say that you always take 100 lengthwise along the

Arithmetic Lines and fit this crosswise to the number of graduations cut by the vertical; then without altering the Instrument you take the distance between points 100–100, which next measured lengthwise will give you the altitude sought. For instance if the thread has cut at 77, take 100 lengthwise along the Arithmetic Lines, fit this crosswise to points 77–77, at once take crosswise the distance between points 100–100, and measure that lengthwise. You will find it to contain 130 graduations, and that many units [of those paced off] you will say to be contained in the height we wished to measure.

We shall be able to measure such a height in another way without the necessity of stepping off 100 units of ground, as will be clearly shown. If, for example, we should wish to measure the height of the tower AB from point C, we could point the side CDE of the Instrument at the summit A, noting the graduation cut by the thread EI, and let this be, say, 80. Then, without changing place, we merely lower the Instrument and sight at some low mark placed on the tower, such as point F, noting the number of the graduation now cut by the thread, say 5. Next see how many times this smaller number, 5, goes into the other, 80, which is 16 times, and we shall say that distance FB is contained 16 times in the whole height BA. And since point F is low, we can easily measure FB with a staff or the like, and thus come to know the altitude BA. Notice that in measuring heights we find and measure only the height above our visual horizon[tal], so that when the eye is above the root or base of the thing being measured one must add, to the height found by using the Instrument, as much more as the eye is above the base.

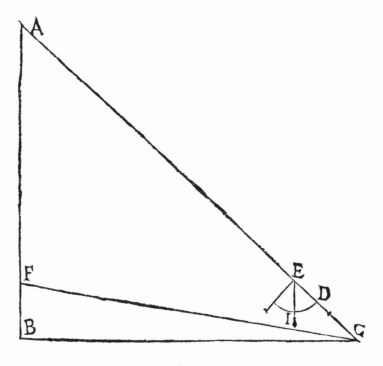

The

The third way to measure such a height will be by rising or descending. Thus, wishing to measure height AB, hold the Instrument at some place above the ground, as at point F, and sight at point A along side EF, noting the graduations between G and I as cut by the thread to be, say, 65. Then, going down and being vertically under point F, say at point C, we sight the same height along side DC, noting the graduations between L and O, which will be more than before, as say 70. Then take the difference between those numbers 65 and 70, which is 5, and as many times as this is contained in the larger number (that is, in 70, which will contain 5 fourteen times), that many times will height BA contain distance CF, which latter we shall measure, as we can easily do, and thus we shall come to know the height of AB.

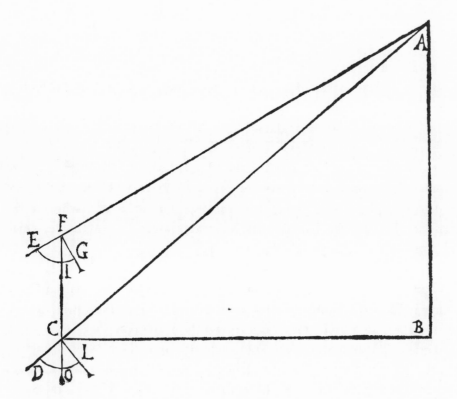

If we want to measure a height whose base cannot be seen, such as the altitude of the mountain AB, we being at point C, let us sight at the summit A and note the graduation at I cut by the vertical DI, and for example let this be 20; next, approaching the mountain 100 paces and coming to point E, we sight the same summit and note graduation F, which is 22. This done, multiply together the numbers 20 and 22, making 440, and divide this by the difference between those same numbers, which is 2, giving 220, and that many paces high we shall say the mountain is.

The

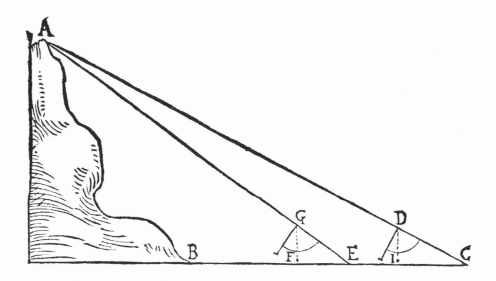

The calculation can be done on the Instrument by taking the smaller number of graduations lengthwise along the Arithmetic Lines and fitting this crosswise at [the numbers representing] the difference between the two numbers of graduations; then take crosswise the larger number of graduations, which measured lengthwise will give us the altitude sought. If, for instance, the graduations cut were 42 and 58, take 42 lengthwise and fit it crosswise to the difference of the said numbers—that is, at 16–16. If unable to do this, use the double, triple, or quadruple; say [you use] the quadruple, which is 64. Then take 58, or [rather] its quadruple 232, and measure this lengthwise, which will give you 152¼, the height sought.

Moreover we can measure with the same Instrument one height placed on top of another, as when we would measure the height of tower AB on top of mountain BC. First, from point D let us sight the top of the tower, A, and note the graduation cut by the thread EI, which for example shall be 18. Then, leaving a staff standing at point D, we advance until the base of the tower, B, is sighted with the vertical GO again cutting the previous number, 18, say when we have got to point F. Then measure the paces between these two stations, D and F, which for example let be 130; multiply that number by the previous 18, making 2,340, which you divide by 100, getting 23⅖, the height of the tower in paces.

On the Instrument the computation will be done by taking lengthwise the number of paces (or that of graduations) and fitting it at points 100–100; then take the other number crosswise and measure this lengthwise. For example if the graduation cut was 64 and the paces were 146, take 64 lengthwise and fit it crosswise to 100–100; then take 146 crosswise and measure this lengthwise, giving us approximately 93½, which is the height sought.

As

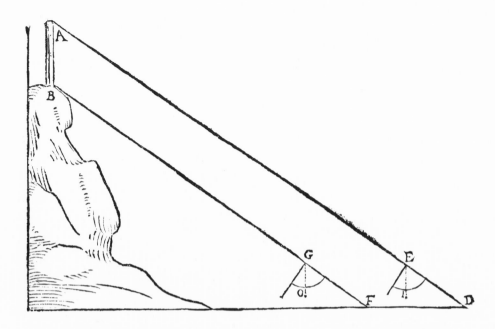

As to depths, we shall have two ways of measuring these. The first will be to measure a depth contained between two parallel lines, like the depth of a well or the [inside] height of a tower from its top. Thus, let there be the well ABDC between the parallel lines AC and DB. With the corner of the Instrument toward the eye E, sight alongside EF in such a way that the visual ray

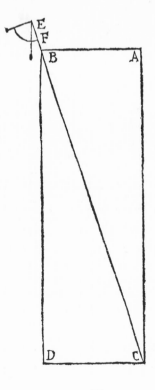

passes

passes through points B and C, noting the number cut by the thread, which for example let be 5. Then see how many times this number goes into 100, and that many times will the breadth BA [which can me measured] be contained in the depth BD.

The other way will be to measure a depth of which the [vertical] base cannot be seen, as when we are on the mountain BA and want to measure its height above the surrounding plain. In such a case we raise ourselves above the mountain, climbing up some house, tower, or tree; and placing the eye at point F we sight at some mark situated in the plain, as at point C, noting the graduation cut by the thread FG, say 32. Then, descending to point D, we sight the same mark C along side DE, noting the graduations along AI, which let be 30. Take the difference between these two numbers, which is 2, and see how many times this goes into the smaller number, which is 15 times; then we say that the height of the mountain is 15 times the height FD, which we can measure, and this will give us what we sought.

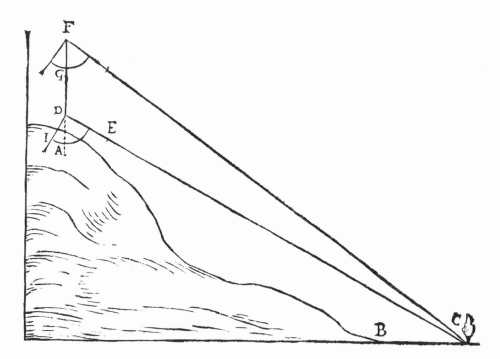

We pass on now to the measurement of distances, such as the width of a river from one of its banks or some other high place, as seen in the diagram. We want to measure the width CB. From point A we sight the [river] edge B along side AF, noting the graduation in DE cut by the vertical, which shall for example be 5. However many times this number goes into 100, that many times will the height AC go into the width CB; so measuring the height AC and multiplying it by 20 we shall have the width sought.

We

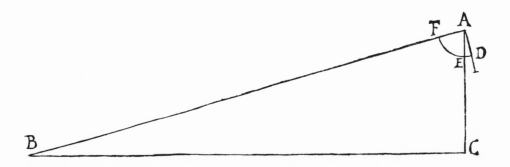

We can measure such distances in another way. Being at point A, for example, we want to find the distance to point B. Set the Instrument at right angles, one leg being pointed toward B, and sight along the other leg toward point C. Measure 100 paces (or other units) in the direction AC and leave a staff driven at point A, placing another at C. Then, at C, point one leg of the Instrument toward A and sight through corner C at mark B, noting on the Quadrant the point cut by the visual ray, which shall be point E. Divide that number into 10,000; what you get will be the number of paces (or other units) between point A and mark B.

But if we are not permitted to move the 100 paces along a line at right angles to the first sighting, we must proceed otherwise. For example we are at point A and want to take the distance D, but cannot move except along AE which makes an acute angle with AB. In order to carry out our intent we shall first aim one leg of our Instrument along this path, as seen by line AF, and without changing that setting we sight point B through angle A, noting the graduation cut by ray AD, which for example is 60. Then, leaving a staff at point A, we shall put another

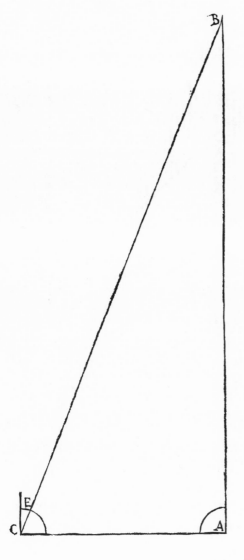

one

one 100 paces along line AE, at point F. There we hold the corner of the Instrument, aiming leg EF at staff A; we sight mark B from corner F, noting the graduations [between] G and I to be, for example, 48. To find distance AB from these numbers 60 and 48, multiply the first by itself, making 3,600; add 10,000 and get 13,600; take the square root of that number, which will be about 117; multiply this by 100 to make 11,700; finally, divide that number by the difference between the two original numbers 60 and 48 (that is, by 12). The result is 975, which is undoubtedly the distance AB, in paces.

Calculation for this operation is performed on the Instrument as explained in the following example. Let the graduations cut by the two visual rays be 74 and 36. To make the computation, first set the Instrument so that the Arithmetic Lines are at right angles, which may be done by taking 100 graduations lengthwise on these and fitting this by a compass across the same Lines in such a way that with one of the [compass] legs at point 80, the other falls at 60 (and this rule for setting the lines at right angles should be kept in mind for other uses). Then take the crosswise distance

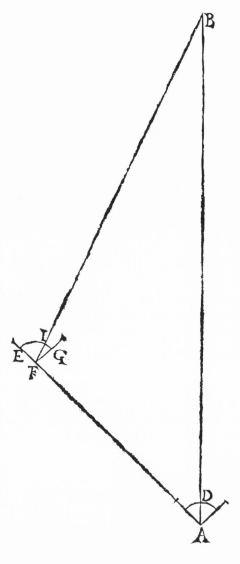

between point 100 and the larger of the two numbers cut by the rays, which was 74; this distance must then be fitted crosswise to the difference between the two numbers of graduations cut by the rays, which is 38. If this cannot be done because of the smallness of the number, then use its double, triple, or quadruple; here, for example, apply its triple, which is 114, and then take the crosswise distance between points 100–100. Measure that distance lengthwise; taken three times, this will be the distance sought. Measured in the present example, you will find it to be 109, so that tripled it will give you 327, which is approximately the distance we wanted to measure.

Next

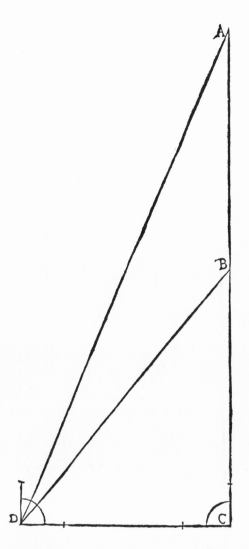

Next let us see how to measure the distance between two places remote from us, and first we shall explain the procedure when from some place we can see them both along the same straight line. This is done in the present example, where we wish to measure the distance between points A and B, which from point C appear along the same line CBA. First set one leg of the Instrument in that direction, and sight [perpendicularly] along the other leg toward D, where a staff is placed 100 units from C, and a similar one at C. Going to point D, we hold one leg of the Instrument in the direction DC, sighting from corner D the two places B and A, and noting the numbers cut by these rays, which are, say, 20 and 25. Divide 10,000 by those numbers, and the difference between the two results will be the distance BA.

But if we want to measure the distance between two places C and D and are unable to reach any place from which they appear in the same direction, we proceed in such cases as next described. Assume that standing at place A, we want to investigate the distance between the two places C and D. First, set one leg of the Instrument as shown by line AEC, sight through an angle the other point, D, and note the graduations [between] E and F (cut by ray AFD), these being for example 20. Without moving the Instrument, sight along the other leg toward B, planting a staff at A and having another one placed along the direction AB. Then, walking in that direction, we get to B, having departed from the second staff just enough so that again pointing one leg of the Instrument along line BA, the other leg points at D (as shown by line BD). From corner B we sight point C, noting the number cut by ray BG, which (say) is 15. Finally, the number of paces between the two stations A and B is measured, which in our example is 160. Now, coming to the arithmetical work, first multiply the number of paces between the two stations (that is, 160)

by

by 100, making 16,000; divide this separately by the numbers of [Quadrant] points—that is, by 20 and by 15—giving us the two numbers 800 and 1,067. Take the difference between these, which is 267, and multiply it by itself, making 71,289. Add that number to the square of the number of paces (160), which is 25,600, for a total of 96,889. Take the square root of this number, which is 311, and that many paces we shall say to lie between the two places C and D.

How one may find the computation on the Instrument will be clear from the following example. Let the two numbers cut by the rays be, for example, 60 and 34, and the number of paces be 116. For the operation, always take 100 lengthwise along the Arithmetic Lines, apply this crosswise at the greater of the numbers cut by the rays (which here is 60), and at once take crosswise the number of paces (here 116). Fit this distance crosswise to the other number of rays, which here is 34; if you cannot, then fit its double, triple, quadruple, or whatever turns out most convenient. Here let us take the quadruple, which is 136. This done, take crosswise the numerical difference between the two numbers cut by the rays, in this case 26, or take its double,

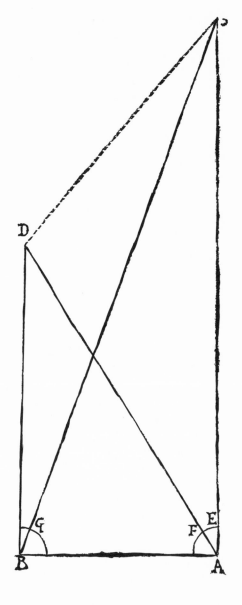

triple, or quadruple according to the application made just previously, whence in this case you should take the quadruple, or 104. Measure this distance lengthwise, keeping in mind the number obtained, which in the present example will be 148. Finally, set the Arithmetic Lines at right angles in the way explained above; that done, take crosswise the difference between the number you kept in memory and the number of paces (that is, between 148 and 116;) measure this lengthwise and you will find 188, which is exactly the distance DC that was sought.

Finally, if we cannot move in the way required in the previous operation, we shall nevertheless be able to find the distance between two remote places in this other way. Being at point C, for example, we wish to find the distance

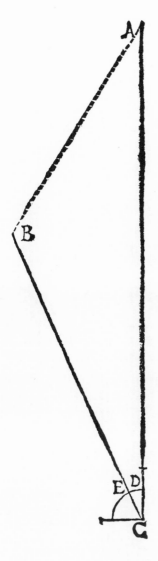

between the two places A and B. By using any of the procedures already explained, we shall measure separately the distance between C and A and the distance between C and B; let the former be for example 850 paces, and the latter 530. At mark C, set one leg of the Instrument to point at A (as shown by line CDA), and from corner C sight at the other terminus B, noting the number of graduations [between] D and E cut by this ray, which for example is 15. Multiply that number by itself, making 225, and 10,000 making 10,225; take the square root, which is 101. Multiply the smaller distance, 530, by 100, making 53,000, which you divide by the root just found, getting 525. This you multiply by the greater distance, 850, making 446,250, which number must be doubled, making 892,500. Next you must multiply separately each of the two distances by itself, getting 722,500 and 280,900; add these together, making 1,003,400. Subtract from that the double found before, 892,500, and there remains 110,900, whose square root, which is 347, will be the distance [in paces] between the two places A and B, which was desired.

With a notable reduction of labor we can make the computation on the Arithmetic Lines in the way made clear by an example. Suppose the longer distance to be 230 paces, the shorter 104, and the number cut by the ray to be 58. Set the Arithmetic Lines at right angles, and placing one leg of a compass at point 100, move the other crosswise to the number of the point cut by the ray, which was 58; then see how much this distance is, measured lengthwise. You will find it to be about 116, to be kept in memory. Then take lengthwise the number 58 (which was the point cut by the ray) and open the Instrument until that distance fits crosswise between 100 [on one arm] and the 116 that you kept in memory [found along the other arm]. Not changing the Instrument, take with a compass the crosswise distance between the two numbers of paces, 230 and 104, and this measured lengthwise finally gives you 150 paces,[24] truly the distance AB.

Gentle Reader, I have judged it sufficient for now to have described these rules for measurement by sighting—not that they alone can be worked on the Instrument, there being a great many others—but in order not to engage unnecessarily in long discourses, I being certain that anyone of average intelligence will have understood those already explained and can find others for himself, suited to any particular case that may arise.

Not only might I have gone on much longer about rules for measurement by sighting, but I could expand on many, many more rules by showing the solution of (I may say) infinitely many other problems of geometry and arithmetic which can be solved with the other Lines of our Instrument; for as many problems as there are in Euclid's *Elements* and in other authors, they are thus resolved by me very quickly and easily. But as was said at the outset, my present intention has been only to speak to military men, and of few things beyond those concerning that profession, reserving to another occasion the publication of the construction of the Instrument and a more ample description of its uses.

THE END.

NOTES

[1]Apparently an allusion to the incident with Jan Eutel Zieckmesser in 1603. This appears to have been the same man as a later mathematician to the Archbishop of Cologne, called Zugmesser, but in the records of the University of Padua the name was spelled as above, as also in Galileo's *Difesa* of 1607.

[2]The Arithmetic Lines and Added Lines were placed on Galileo's instrument shortly before March 1599.

[3]Vincenzo Gonzaga, with whom Galileo negotiated for a court post in 1603–04; at that time he aided Aurelio Capra by recommending him to the Duke in the matter of a medical secret.

[4]The word "compass" is used throughout this book in the sense of "pair of dividers"; the "military compass" is called by Galileo "the Instrument." Galileo's use of capitals and numerals has been slightly altered in the interest of clarity and ease of reading. Only enough capitals have been retained, and numbers written out, to preserve the flavor of the original.

[5]We would say that AC is seven times AB; Galileo often used "n times greater than" to mean "$(n + 1)$ times."

[6]The plural "parts" means non-aliquot parts, the singular "part" meaning aliquot part; cf. Euclid, Book VII, Defs. 3 and 4.

[7]A measure of area.

[8]In the manuscripts circulated from 1600 to 1605 the word Chapter was used, rather than Operation as in the printed text. Occurrence of "Chapter" in this text probably marks a section of text left unrevised in Galileo's final editing.

[9]The ratios would be the same as for diameters, but radii are more nearly analogous to the sides of rectilineal figures.

[10]"Geometrically double" means to quadruple; "geometrically multiply" means to multiply by the square of the denominator of the unit fraction named.

[11]The original text read "Geometric," corrected in some copies in Galileo's handwriting.

[12]That is, $ax^2/b = c$. At this period separate rules were used for different types of quadratic equations, classified according to the presence or absence of a term in x and the positions of all terms to right or left of the $=$ sign. Galileo never used algebra, although he was familiar with its use in solving numerical problems; this may be the only place he mentioned the word in print.

[13]That is, must be multiplied by 2^3, or 8; cf. note 10, above.

[14]The diagram as given in the National Edition is not exactly as in the original edition.

[15]By taking any point inside the figure and connecting this to each of the angles. The previous constructions were related to regular polygons only, which would include only equilateral triangles, whereas the problem here is to square any triangle.

[16]Called by Galileo "compass of four points." See Introduction.

[17]This curious marking is not found on any actual instrument. Possibly the printer intended it to be a square, made up of eight small square dots, which became disarranged in the printing. Not all copies of the 1606 edition show the mark as reproduced in the National Edition. The existing Paduan instruments have no marking at all, and none is really needed, as the intersection of the Added Line with the ruling along the outer edge provides the basis of measurement.

[18]The device introduced by Tartaglia and in use throughout Europe; see Introduction.

[19]On Galileo's own instrument a projection at the zero point of this scale replaced the hole described in the text.

[20]The *braccio* (plural, *braccia*) was much used as a measure of length, but differed from one city to another. The Florentine braccio was about 23 inches as shown in Galileo's syllabus on fortification written at Padua.

[21]From ancient Roman times this had been the usual way of describing slopes. Galileo's slanting lines advanced by units we would call "100 percent of grade."

[22]The arc to the right of I in the diagram is meant. This being less than 45° from the vertical, it would fall in the scale of 100 graduations that were more distant from the eye.

[23]Explanations of simple numerical calculations were included, for the benefit of uneducated practical men, in Galileo's manuscript versions from 1599 on. These were partly abbreviated and partly omitted in the printed book.

[24]Instead of "paces," the text reads "points," that being the word by which Galileo named the distance (about 29/30 mm.) between adjacent markings along the

Arithmetic Line. It is of interest that, in this concluding section, Galileo took cross-wise distances between different points along the same scales, rather than between like points only as had been done in all earlier sections. This technique greatly shortens some calculations by treating the sector much as if it were an analogue computer, creating scale models of the large triangles that were to be analyzed. No such procedures were possible on the slide rule or the calculating machine, which may explain why the sector continued in use long after their invention.